HOLT SCIENCE & TECHNOLOGY

Weather and Climate

HOLT, RINEHART AND WINSTON

A Harcourt Classroom Education Company

Austin • New York • Orlando • Atlanta • San Francisco • Boston • Dallas • Toronto • London

Staff Credits

Editorial

Robert W. Todd, Executive Editor
Robert V. Tucek, Leigh Ann Garcia, Senior Editors
Clay Walton, Jim Ratcliffe, Editors

ANCILLARIES

Jennifer Childers, Senior Editor
Chris Colby, Molly Frohlich, Shari Husain, Kristen McCardel, Sabelyn Pussman, Erin Roberson

COPYEDITING

Dawn Spinozza, Copyediting Supervisor

EDITORIAL SUPPORT STAFF

Jeanne Graham, Mary Helbling, Tanu'e White, Doug Rutley

EDITORIAL PERMISSIONS

Cathy Paré, Permissions Manager
Jan Harrington, Permissions Editor

Art, Design, and Photo

BOOK DESIGN

Richard Metzger, Design Director
Marc Cooper, Senior Designer
José Garza, Designer
Alicia Sullivan, Designer (ATE), **Cristina Bowerman**, Design Associate (ATE), **Eric Rupprath**, Designer (Ancillaries), **Holly Whittaker**, Traffic Coordinator

IMAGE ACQUISITIONS

Joe London, Director
Elaine Tate, Art Buyer Supervisor
Jeannie Taylor, Photo Research Supervisor
Andy Christiansen, Photo Researcher
Jackie Berger, Assistant Photo Researcher

PHOTO STUDIO

Sam Dudgeon, Senior Staff Photographer
Victoria Smith, Photo Specialist
Lauren Eischen, Photo Coordinator

DESIGN NEW MEDIA

Susan Michael, Design Director

Production

Mimi Stockdell, Senior Production Manager
Beth Sample, Senior Production Coordinator
Suzanne Brooks, Sara Carroll-Downs

Media Production

Kim A. Scott, Senior Production Manager
Adriana Bardin-Prestwood, Senior Production Coordinator

New Media

Armin Gutzmer, Director
Jim Bruno, Senior Project Manager
Lydia Doty, Senior Project Manager
Jessica Bega, Project Manager
Cathy Kuhles, Nina Degollado, Technical Assistants

Design Implementation and Production

The Quarasan Group, Inc.

Acknowledgments

Chapter Writers

Kathleen Meehan Berry
Science Chairman
Canon-McMillan School
  District
Canonsburg, Pennsylvania

Robert H. Fronk, Ph.D.
Chair of Science and Mathematics
  Education Department
Florida Institute of Technology
West Melbourne, Florida

Mary Kay Hemenway, Ph.D.
Research Associate and Senior
  Lecturer
Department of Astronomy
The University of Texas
Austin, Texas

Kathleen Kaska
Life and Earth Science Teacher
Lake Travis Middle School
Austin, Texas

Peter E. Malin, Ph.D.
Professor of Geology
Division of Earth and Ocean
  Sciences
Duke University
Durham, North Carolina

Karen J. Meech, Ph.D.
Associate Astronomer
Institute for Astronomy
University of Hawaii
Honolulu, Hawaii

Robert J. Sager
Chair and Professor of Earth
  Sciences
Pierce College
Lakewood, Washington

Lab Writers

Kenneth Creese
Science Teacher
White Mountain Junior
  High School
Rock Springs, Wyoming

Linda A. Culp
Science Teacher and Dept. Chair
Thorndale High School
Thorndale, Texas

Bruce M. Jones
Science Teacher and Dept. Chair
The Blake School
Minneapolis, Minnesota

Shannon Miller
Science and Math Teacher
Llano Junior High School
Llano, Texas

Robert Stephen Ricks
Special Services Teacher
Department of Classroom
  Improvement
Alabama State Department
  of Education
Montgomery, Alabama

James J. Secosky
Science Teacher
Bloomfield Central School
Bloomfield, New York

Academic Reviewers

Mead Allison, Ph.D.
Assistant Professor of
  Oceanography
Texas A&M University
Galveston, Texas

Alissa Arp, Ph.D.
Director and Professor of
  Environmental Studies
Romberg Tiburon Center
San Francisco State University
Tiburon, California

Paul D. Asimow, Ph.D.
Assistant Professor of Geology and
  Geochemistry
Department of Physics and
  Planetary Sciences
California Institute of
  Technology
Pasadena, California

G. Fritz Benedict, Ph.D.
Senior Research Scientist and
  Astronomer
McDonald Observatory
The University of Texas
Austin, Texas

Russell M. Brengelman, Ph.D.
Professor of Physics
Morehead State University
Morehead, Kentucky

John A. Brockhaus, Ph.D.
Director—Mapping, Charting, and
  Geodesy Program
Department of Geography and
  Environmental Engineering
United States Military Academy
West Point, New York

Michael Brown, Ph.D.
Assistant Professor of Planetary
  Astronomy
Department of Physics
  and Astronomy
California Institute of
  Technology
Pasadena, California

Wesley N. Colley, Ph.D.
Postdoctoral Fellow
Harvard-Smithsonian Center
  for Astrophysics
Cambridge, Massachusetts

Andrew J. Davis, Ph.D.
Manager—ACE Science Data
  Center
Physics Department
California Institute of
  Technology
Pasadena, California

Peter E. Demmin, Ed.D.
Former Science Teacher and
  Department Chair
Amherst Central High School
Amherst, New York

James Denbow, Ph.D.
Associate Professor
Department of Anthropology
The University of Texas
Austin, Texas

Roy W. Hann, Jr., Ph.D.
Professor of Civil Engineering
Texas A&M University
College Station, Texas

Frederick R. Heck, Ph.D.
Professor of Geology
Ferris State University
Big Rapids, Michigan

Richard Hey, Ph.D.
Professor of Geophysics
Hawaii Institute of Geophysics
  and Planetology
University of Hawaii
Honolulu, Hawaii

John E. Hoover, Ph.D.
Associate Professor of Biology
Millersville University
Millersville, Pennsylvania

Robert W. Houghton, Ph.D.
Senior Staff Associate
Lamont-Doherty Earth
  Observatory
Columbia University
Palisades, New York

Steven A. Jennings, Ph.D.
Assistant Professor
Department of Geography &
  Environmental Studies
University of Colorado
Colorado Springs, Colorado

Eric L. Johnson, Ph.D.
Assistant Professor of Geology
Central Michigan University
Mount Pleasant, Michigan

John Kermond, Ph.D.
Visiting Scientist
NOAA–Office of Global
  Programs
Silver Spring, Maryland

Zavareh Kothavala, Ph.D.
Postdoctoral Associate Scientist
Department of Geology and
  Geophysics
Yale University
New Haven, Connecticut

Karen Kwitter, Ph.D.
Ebenezer Fitch Professor of
  Astronomy
Williams College
Williamstown, Massachusetts

Valerie Lang, Ph.D.
Project Leader of Environmental
  Programs
The Aerospace Corporation
Los Angeles, California

Philip LaRoe
Professor
Helena College of Technology
Helena, Montana

Julie Lutz, Ph.D.
Astronomy Program
Washington State University
Pullman, Washington

Duane F. Marble, Ph.D.
Professor Emeritus
Department of Geography and
  Natural Resources
Ohio State University
Columbus, Ohio

Joseph A. McClure, Ph.D.
Associate Professor
Department of Physics
Georgetown University
Washington, D.C.

Frank K. McKinney, Ph.D.
Professor of Geology
Appalachian State University
Boone, North Carolina

Joann Mossa, Ph.D.
Associate Professor of Geography
University of Florida
Gainesville, Florida

LaMoine L. Motz, Ph.D.
Coordinator of Science Education
Department of Learning
  Services
Oakland County Schools
Waterford, Michigan

Barbara Murck, Ph.D.
Assistant Professor of Earth
  Science
Erindale College
University of Toronto
Mississauga, Ontario,
  Canada

Hilary Clement Olson, Ph.D.
Research Associate
Institute for Geophysics
The University of Texas
Austin, Texas

Andre Potochnik
Geologist
Grand Canyon Field Institute
Flagstaff, Arizona

John R. Reid, Ph.D.
Professor Emeritus
Department of Geology and
  Geological Engineering
University of North Dakota
Grand Forks, North Dakota

Gary Rottman, Ph.D.
Associate Director
Laboratory for Atmosphere and
  Space Physics
University of Colorado
Boulder, Colorado

Dork L. Sahagian, Ph.D.
Professor
Institute for the Study of Earth,
  Oceans, and Space
University of New Hampshire
Durham, New Hampshire

Peter Sheridan, Ph.D.
Professor of Chemistry
Colgate University
Hamilton, New York

David Sprayberry, Ph.D.
Assistant Director for Observing
  Support
W.M. Keck Observatory
California Association for
  Research in Astronomy
Kamuela, Hawaii

Lynne Talley, Ph.D.
Professor
Scripps Institution of
  Oceanography
University of California
La Jolla, California

T3

Acknowledgments (cont.)

Glenn Thompson, Ph.D.
Scientist
Geophysical Institute
University of Alaska
Fairbanks, Alaska

Martin VanDyke, Ph.D.
Professor of Chemistry, Emeritus
Front Range Community
    College
Westminister, Colorado

Thad A. Wasklewicz, Ph.D.
Assistant Professor of Geography
University of Memphis
Memphis, Tennessee

Hans Rudolf Wenk, Ph.D.
*Professor of Geology and
    Geophysical Sciences*
University of California
Berkeley, California

Lisa D. White, Ph.D.
Associate Professor of Geosciences
San Francisco State University
San Francisco, California

Lorraine W. Wolf, Ph.D.
Associate Professor of Geology
Auburn University
Auburn, Alabama

Charles A. Wood, Ph.D.
*Chairman and Professor of Space
    Studies*
University of North Dakota
Grand Forks, North Dakota

Safety Reviewer

Jack Gerlovich, Ph.D.
Associate Professor
School of Education
Drake University
Des Moines, Iowa

Teacher Reviewers

Barry L. Bishop
Science Teacher and Dept. Chair
San Rafael Junior High School
Ferron, Utah

Yvonne Brannum
Science Teacher and Dept. Chair
Hine Junior High School
Washington, D.C.

Daniel L. Bugenhagen
Science Teacher and Dept. Chair
Yutan Junior & Senior High
    School
Yutan, Nebraska

Kenneth Creese
Science Teacher
White Mountain Junior High
    School
Rock Springs, Wyoming

Linda A. Culp
Science Teacher and Dept. Chair
Thorndale High School
Thorndale, Texas

Alonda Droege
Science Teacher
Pioneer Middle School
Steilacom, Washington

Laura Fleet
Science Teacher
Alice B. Landrum Middle
    School
Ponte Vedra Beach, Florida

Susan Gorman
Science Teacher
Northridge Middle School
North Richland Hills, Texas

C. John Graves
Science Teacher
Monforton Middle School
Bozeman, Montana

Janel Guse
Science Teacher and Dept. Chair
West Central Middle School
Hartford, South Dakota

Gary Habeeb
Science Mentor
Sierra–Plumas Joint Unified
    School District
Downieville, California

Dennis Hanson
Science Teacher and Dept. Chair
Big Bear Middle School
Big Bear Lake, California

Norman E. Holcomb
Science Teacher
Marion Local Schools
Maria Stein, Ohio

Tracy Jahn
Science Teacher
Berkshire Junior-Senior High
    School
Canaan, New York

David D. Jones
Science Teacher
Andrew Jackson Middle School
Cross Lanes, West Virginia

Howard A. Knodle
Science Teacher
Belvidere High School
Belvidere, Illinois

Michael E. Kral
Science Teacher
West Hardin Middle School
Cecilia, Kentucky

Kathy LaRoe
Science Teacher
East Valley Middle School
East Helena, Montana

Scott Mandel, Ph.D.
*Director and Educational
    Consultant*
Teachers Helping Teachers
Los Angeles, California

Kathy McKee
Science Teacher
Hoyt Middle School
Des Moines, Iowa

Michael Minium
*Vice President of Program
    Development*
United States Orienteering
    Federation
Forest Park, Georgia

Jan Nelson
Science Teacher
East Valley Middle School
East Helena, Montana

Dwight C. Patton
Science Teacher
Carroll T. Welch Middle School
Horizon City, Texas

Joseph Price
Chairman—Science Department
H. M. Brown Junior High
    School
Washington, D.C.

Terry J. Rakes
Science Teacher
Elmwood Junior High School
Rogers, Arkansas

Steven Ramig
Science Teacher
West Point High School
West Point, Nebraska

Helen P. Schiller
Science Teacher
Northwood Middle School
Taylors, South Carolina

Bert J. Sherwood
Science Teacher
Socorro Middle School
El Paso, Texas

Larry Tackett
Science Teacher and Dept. Chair
Andrew Jackson Middle School
Cross Lanes, West Virginia

Walter Woolbaugh
Science Teacher
Manhattan Junior High School
Manhattan, Montana

Alexis S. Wright
Middle School Science Coordinator
Rye Country Day School
Rye, New York

Gordon Zibelman
Science Teacher
Drexel Hill Middle School
Drexel Hill, Pennsylvania

Weather and Climate

Skills Development

Process Skills

QuickLabs

Chapter Labs

Research and Critical Thinking Skills

Apply

Feature Articles

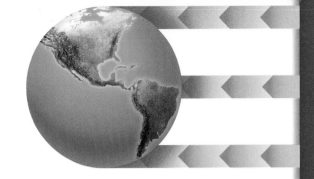

Connections

Program Scope and Sequence

Selecting the right books for your course is easy. Just review the topics presented in each book to determine the best match to your district curriculum.

	A MICROORGANISMS, FUNGI, AND PLANTS	**B** ANIMALS	
CHAPTER 1	**It's Alive!! Or, Is It?** ❑ Characteristics of living things ❑ Homeostasis ❑ Heredity and DNA ❑ Producers, consumers, and decomposers ❑ Biomolecules	**Animals and Behavior** ❑ Characteristics of animals ❑ Classification of animals ❑ Animal behavior ❑ Hibernation and estivation ❑ The biological clock ❑ Animal communication ❑ Living in groups	
CHAPTER 2	**Bacteria and Viruses** ❑ Binary fission ❑ Characteristics of bacteria ❑ Nitrogen-fixing bacteria ❑ Antibiotics ❑ Pathogenic bacteria ❑ Characteristics of viruses ❑ Lytic cycle	**Invertebrates** ❑ General characteristics of invertebrates ❑ Types of symmetry ❑ Characteristics of sponges, cnidarians, arthropods, and echinoderms ❑ Flatworms versus roundworms ❑ Types of circulatory systems	
CHAPTER 3	**Protists and Fungi** ❑ Characteristics of protists ❑ Types of algae ❑ Types of protozoa ❑ Protist reproduction ❑ Characteristics of fungi and lichens	**Fishes, Amphibians, and Reptiles** ❑ Characteristics of vertebrates ❑ Structure and kinds of fishes ❑ Development of lungs ❑ Structure and kinds of amphibians and reptiles ❑ Function of the amniotic egg	
CHAPTER 4	**Introduction to Plants** ❑ Characteristics of plants and seeds ❑ Reproduction and classification ❑ Angiosperms versus gymnosperms ❑ Monocots versus dicots ❑ Structure and functions of roots, stems, leaves, and flowers	**Birds and Mammals** ❑ Structure and kinds of birds ❑ Types of feathers ❑ Adaptations for flight ❑ Structure and kinds of mammals ❑ Function of the placenta	
CHAPTER 5	**Plant Processes** ❑ Pollination and fertilization ❑ Dormancy ❑ Photosynthesis ❑ Plant tropisms ❑ Seasonal responses of plants		
CHAPTER 6			
CHAPTER 7			

Life Science

C CELLS, HEREDITY, & CLASSIFICATION

Cells: The Basic Units of Life
- ❏ Cells, tissues, and organs
- ❏ Populations, communities, and ecosystems
- ❏ Cell theory
- ❏ Surface-to-volume ratio
- ❏ Prokaryotic versus eukaryotic cells
- ❏ Cell organelles

The Cell in Action
- ❏ Diffusion and osmosis
- ❏ Passive versus active transport
- ❏ Endocytosis versus exocytosis
- ❏ Photosynthesis
- ❏ Cellular respiration and fermentation
- ❏ Cell cycle

Heredity
- ❏ Dominant versus recessive traits
- ❏ Genes and alleles
- ❏ Genotype, phenotype, the Punnett square and probability
- ❏ Meiosis
- ❏ Determination of sex

Genes and Gene Technology
- ❏ Structure of DNA
- ❏ Protein synthesis
- ❏ Mutations
- ❏ Heredity disorders and genetic counseling

The Evolution of Living Things
- ❏ Adaptations and species
- ❏ Evidence for evolution
- ❏ Darwin's work and natural selection
- ❏ Formation of new species

The History of Life on Earth
- ❏ Geologic time scale and extinctions
- ❏ Plate tectonics
- ❏ Human evolution

Classification
- ❏ Levels of classification
- ❏ Cladistic diagrams
- ❏ Dichotomous keys
- ❏ Characteristics of the six kingdoms

D HUMAN BODY SYSTEMS & HEALTH

Body Organization and Structure
- ❏ Homeostasis
- ❏ Types of tissue
- ❏ Organ systems
- ❏ Structure and function of the skeletal system, muscular system, and integumentary system

Circulation and Respiration
- ❏ Structure and function of the cardiovascular system, lymphatic system, and respiratory system
- ❏ Respiratory disorders

The Digestive and Urinary Systems
- ❏ Structure and function of the digestive system
- ❏ Structure and function of the urinary system

Communication and Control
- ❏ Structure and function of the nervous system and endocrine system
- ❏ The senses
- ❏ Structure and function of the eye and ear

Reproduction and Development
- ❏ Asexual versus sexual reproduction
- ❏ Internal versus external fertilization
- ❏ Structure and function of the human male and female reproductive systems
- ❏ Fertilization, placental development, and embryo growth
- ❏ Stages of human life

Body Defenses and Disease
- ❏ Types of diseases
- ❏ Vaccines and immunity
- ❏ Structure and function of the immune system
- ❏ Autoimmune diseases, cancer, and AIDS

Staying Healthy
- ❏ Nutrition and reading food labels
- ❏ Alcohol and drug effects on the body
- ❏ Hygiene, exercise, and first aid

E ENVIRONMENTAL SCIENCE

Interactions of Living Things
- ❏ Biotic versus abiotic parts of the environment
- ❏ Producers, consumers, and decomposers
- ❏ Food chains and food webs
- ❏ Factors limiting population growth
- ❏ Predator-prey relationships
- ❏ Symbiosis and coevolution

Cycles in Nature
- ❏ Water cycle
- ❏ Carbon cycle
- ❏ Nitrogen cycle
- ❏ Ecological succession

The Earth's Ecosystems
- ❏ Kinds of land and water biomes
- ❏ Marine ecosystems
- ❏ Freshwater ecosystems

Environmental Problems and Solutions
- ❏ Types of pollutants
- ❏ Types of resources
- ❏ Conservation practices
- ❏ Species protection

Energy Resources
- ❏ Types of resources
- ❏ Energy resources and pollution
- ❏ Alternative energy resources

	F INSIDE THE RESTLESS EARTH	**G** EARTH'S CHANGING SURFACE	
CHAPTER 1	**Minerals of the Earth's Crust** ❏ Mineral composition and structure ❏ Types of minerals ❏ Mineral identification ❏ Mineral formation and mining	**Maps as Models of the Earth** ❏ Structure of a map ❏ Cardinal directions ❏ Latitude, longitude, and the equator ❏ Magnetic declination and true north ❏ Types of projections ❏ Aerial photographs ❏ Remote sensing ❏ Topographic maps	
CHAPTER 2	**Rocks: Mineral Mixtures** ❏ Rock cycle and types of rocks ❏ Rock classification ❏ Characteristics of igneous, sedimentary, and metamorphic rocks	**Weathering and Soil Formation** ❏ Types of weathering ❏ Factors affecting the rate of weathering ❏ Composition of soil ❏ Soil conservation and erosion prevention	
CHAPTER 3	**The Rock and Fossil Record** ❏ Uniformitarianism versus catastrophism ❏ Superposition ❏ The geologic column and unconformities ❏ Absolute dating and radiometric dating ❏ Characteristics and types of fossils ❏ Geologic time scale	**Agents of Erosion and Deposition** ❏ Shoreline erosion and deposition ❏ Wind erosion and deposition ❏ Erosion and deposition by ice ❏ Gravity's effect on erosion and deposition	
CHAPTER 4	**Plate Tectonics** ❏ Structure of the Earth ❏ Continental drifts and sea floor spreading ❏ Plate tectonics theory ❏ Types of boundaries ❏ Types of crust deformities		
CHAPTER 5	**Earthquakes** ❏ Seismology ❏ Features of earthquakes ❏ P and S waves ❏ Gap hypothesis ❏ Earthquake safety		
CHAPTER 6	**Volcanoes** ❏ Types of volcanoes and eruptions ❏ Types of lava and pyroclastic material ❏ Craters versus calderas ❏ Sites and conditions for volcano formation ❏ Predicting eruptions		

Earth Science

Scope and Sequence (continued)

	K INTRODUCTION TO MATTER	**L INTERACTIONS OF MATTER**	
CHAPTER 1	**The Properties of Matter** ❏ Definition of matter ❏ Mass and weight ❏ Physical and chemical properties ❏ Physical and chemical change ❏ Density	**Chemical Bonding** ❏ Types of chemical bonds ❏ Valence electrons ❏ Ions versus molecules ❏ Crystal lattice	
CHAPTER 2	**States of Matter** ❏ States of matter and their properties ❏ Boyle's and Charles's laws ❏ Changes of state	**Chemical Reactions** ❏ Writing chemical formulas and equations ❏ Law of conservation of mass ❏ Types of reactions ❏ Endothermic versus exothermic reactions ❏ Law of conservation of energy ❏ Activation energy ❏ Catalysts and inhibitors	
CHAPTER 3	**Elements, Compounds, and Mixtures** ❏ Elements and compounds ❏ Metals, nonmetals, and metalloids (semiconductors) ❏ Properties of mixtures ❏ Properties of solutions, suspensions, and colloids	**Chemical Compounds** ❏ Ionic versus covalent compounds ❏ Acids, bases, and salts ❏ pH ❏ Organic compounds ❏ Biomolecules	
CHAPTER 4	**Introduction to Atoms** ❏ Atomic theory ❏ Atomic model and structure ❏ Isotopes ❏ Atomic mass and mass number	**Atomic Energy** ❏ Properties of radioactive substances ❏ Types of decay ❏ Half-life ❏ Fission, fusion, and chain reactions	
CHAPTER 5	**The Periodic Table** ❏ Structure of the periodic table ❏ Periodic law ❏ Properties of alkali metals, alkaline-earth metals, halogens, and noble gases		
CHAPTER 6			

Physical Science

Matter in Motion
- ❏ Speed, velocity, and acceleration
- ❏ Measuring force
- ❏ Friction
- ❏ Mass versus weight

Forces in Motion
- ❏ Terminal velocity and free fall
- ❏ Projectile motion
- ❏ Inertia
- ❏ Momentum

Forces in Fluids
- ❏ Properties in fluids
- ❏ Atmospheric pressure
- ❏ Density
- ❏ Pascal's principle
- ❏ Buoyant force
- ❏ Archimedes' principle
- ❏ Bernoulli's principle

Work and Machines
- ❏ Measuring work
- ❏ Measuring power
- ❏ Types of machines
- ❏ Mechanical advantage
- ❏ Mechanical efficiency

Energy and Energy Resources
- ❏ Forms of energy
- ❏ Energy conversions
- ❏ Law of conservation of energy
- ❏ Energy resources

Heat and Heat Technology
- ❏ Heat versus temperature
- ❏ Thermal expansion
- ❏ Absolute zero
- ❏ Conduction, convection, radiation
- ❏ Conductors versus insulators
- ❏ Specific heat capacity
- ❏ Changes of state
- ❏ Heat engines
- ❏ Thermal pollution

Introduction to Electricity
- ❏ Law of electric charges
- ❏ Conduction versus induction
- ❏ Static electricity
- ❏ Potential difference
- ❏ Cells, batteries, and photocells
- ❏ Thermocouples
- ❏ Voltage, current, and resistance
- ❏ Electric power
- ❏ Types of circuits

Electromagnetism
- ❏ Properties of magnets
- ❏ Magnetic force
- ❏ Electromagnetism
- ❏ Solenoids and electric motors
- ❏ Electromagnetic induction
- ❏ Generators and transformers

Electronic Technology
- ❏ Properties of semiconductors
- ❏ Integrated circuits
- ❏ Diodes and transistors
- ❏ Analog versus digital signals
- ❏ Microprocessors
- ❏ Features of computers

The Energy of Waves
- ❏ Properties of waves
- ❏ Types of waves
- ❏ Reflection and refraction
- ❏ Diffraction and interference
- ❏ Standing waves and resonance

The Nature of Sound
- ❏ Properties of sound waves
- ❏ Structure of the human ear
- ❏ Pitch and the Doppler effect
- ❏ Infrasonic versus ultrasonic sound
- ❏ Sound reflection and echolocation
- ❏ Sound barrier
- ❏ Interference, resonance, diffraction, and standing waves
- ❏ Sound quality of instruments

The Nature of Light
- ❏ Electromagnetic waves
- ❏ Electromagnetic spectrum
- ❏ Law of reflection
- ❏ Absorption and scattering
- ❏ Reflection and refraction
- ❏ Diffraction and interference

Light and Our World
- ❏ Luminosity
- ❏ Types of lighting
- ❏ Types of mirrors and lenses
- ❏ Focal point
- ❏ Structure of the human eye
- ❏ Lasers and holograms

HOLT SCIENCE & TECHNOLOGY

Components Listing

Effective planning starts with all the resources you need in an easy-to-use package for each short course.

Directed Reading Worksheets Help students develop and practice fundamental reading comprehension skills and provide a comprehensive review tool for students to use when studying for an exam.

Study Guide Vocabulary & Notes Worksheets and Chapter Review Worksheets are reproductions of the Chapter Highlights and Chapter Review sections that follow each chapter in the textbook.

Science Puzzlers, Twisters & Teasers Use vocabulary and concepts from each chapter of the Pupil's Editions as elements of rebuses, anagrams, logic puzzles, daffy definitions, riddle poems, word jumbles, and other types of puzzles.

Reinforcement and Vocabulary Review Worksheets Approach a chapter topic from a different angle with an emphasis on different learning modalities to help students that are frustrated by traditional methods.

Critical Thinking & Problem Solving Worksheets Develop the following skills: distinguishing fact from opinion, predicting consequences, analyzing information, and drawing conclusions. Problem Solving Worksheets develop a step-by-step process of problem analysis including gathering information, asking critical questions, identifying alternatives, and making comparisons.

Math Skills for Science Worksheets Each activity gives a brief introduction to a relevant math skill, a step-by-step explanation of the math process, one or more example problems, and a variety of practice problems.

Science Skills Worksheets Help your students focus specifically on skills such as measuring, graphing, using logic, understanding statistics, organizing research papers, and critical thinking options.

LAB ACTIVITIES

ALL LABS ARE CLASSROOM TESTED & APPROVED

Datasheets for Labs These worksheets are the labs found in the *Holt Science & Technology* textbook. Charts, tables, and graphs are included to make data collection and analysis easier, and space is provided to write observations and conclusions.

Whiz-Bang Demonstrations Discovery or Making Models experiences label each demo as one in which students discover an answer or use a scientific model.

Calculator-Based Labs Give students the opportunity to use graphing-calculator probes and sensors to collect data using a TI graphing calculator, Vernier sensors, and a TI CBL 2™ or Vernier Lab Pro interface.

EcoLabs and Field Activities Focus on educational outdoor projects, such as wildlife observation, nature surveys, or natural history.

Inquiry Labs Use the scientific method to help students find their own path in solving a real-world problem.

Long-Term Projects and Research Ideas Provide students with the opportunity to go beyond library and Internet resources to explore science topics.

ASSESSMENT

Chapter Tests Each four-page chapter test consists of a variety of item types including Multiple Choice, Using Vocabulary, Short Answer, Critical Thinking, Math in Science, Interpreting Graphics, and Concept Mapping.

Performance-Based Assessments Evaluate students' abilities to solve problems using the tools, equipment, and techniques of science. Rubrics included for each assessment make it easy to evaluate student performance.

TEACHER RESOURCES

Lesson Plans Integrate all of the great resources in the *Holt Science & Technology* program into your daily teaching. Each lesson plan includes a correlation of the lesson activities to the National Science Education Standards.

Teaching Transparencies Each transparency is correlated to a particular lesson in the Chapter Organizer.

 **Concept Mapping Transparencies, Worksheets, and Answer Key**

Give students an opportunity to complete their own concept maps to study the concepts within each chapter and form logical connections. Student worksheets contain a blank concept map with linking phrases and a list of terms to be used by the student to complete the map.

TECHNOLOGY RESOURCES

One-Stop Planner CD-ROM

Finding the right resources is easy with the One-Stop Planner CD-ROM. You can view and print any resource with just the click of a mouse. Customize the suggested lesson plans to match your daily or weekly calendar and your district's requirements. Powerful test generator software allows you to create customized assessments using a databank of items.

The One-Stop Planner for each level includes the following:

- All materials from the Teaching Resources
- Bellringer Transparency Masters
- Block Scheduling Tools
- Standards Correlations
- Lab Inventory Checklist
- Safety Information
- Science Fair Guide
- Parent Involvement Tools
- Spanish Audio Scripts
- Spanish Glossary
- Assessment Item Listing
- Assessment Checklists and Rubrics
- Test Generator

sciLINKS

*sci*LINKS numbers throughout the text take you and your students to some of the best on-line resources available. Sites are constantly reviewed and updated by the National Science Teachers Association. Special "teacher only" sites are available to you once you register with the service.

go.hrw.com

To access Holt, Rinehart and Winston Web resources, use the home page codes for each level found on page 1 of the Pupil's Editions. The codes shown on the Chapter Organizers for each chapter in the Annotated Teacher's Edition take you to chapter-specific resources.

Smithsonian Institution

Find lesson plans, activities, interviews, virtual exhibits, and just general information on a wide variety of topics relevant to middle school science.

CNNfyi.com

Find the latest in late-breaking science news for students. Featured news stories are supported with lesson plans and activities.

CNN Presents Science in the News Video Library

Bring relevant science news stories into the classroom. Each video comes with a Teacher's Guide and set of Critical Thinking Worksheets that develop listening and media analysis skills. Tapes in the series include:

- Eye on the Environment
- Multicultural Connections
- Scientists in Action
- Science, Technology & Society

Guided Reading Audio CD Program

Students can listen to a direct read of each chapter and follow along in the text. Use the program as a content bridge for struggling readers and students for whom English is not their native language.

Interactive Explorations CD-ROM

Turn a computer into a virtual laboratory. Students act as lab assistants helping Dr. Crystal Labcoat solve real-world problems. Activities develop students' inquiry, analysis, and decision-making skills.

Interactive Science Encyclopedia CD-ROM

Give your students access to more than 3,000 cross-referenced scientific definitions, in-depth articles, science fair project ideas, activities, and more.

ADDITIONAL COMPONENTS

Holt Anthology of Science Fiction

Science Fiction features in the Pupil's Edition preview the stories found in the anthology. Each story begins with a Reading Prep guide and closes with Think About It questions.

Professional Reference for Teachers

Articles written by leading educators help you learn more about the National Science Education Standards, block scheduling, classroom management techniques, and more. A bibliography of professional references is included.

Holt Science Posters

Seven wall posters highlight interesting topics, such as the Physics of Sports, or useful reference material, such as the Scientific Method.

 Holt Science Skills Workshop: Reading in the Content Area

Use a variety of in-depth skills exercises to help students learn to read science materials strategically.

Key
These materials are blackline masters.
All titles shown in green are found in the *Teaching Resources* booklets for each course.

Science & Math Skills Worksheets

The *Holt Science and Technology* program helps you meet the needs of a wide variety of students, regardless of their skill level. The following pages provide examples of the worksheets available to improve your students' science and math skills, whether they already have a strong science and math background or are weak in these areas. Samples of assessment checklists and rubrics are also provided.

In addition to the skills worksheets represented here, *Holt Science and Technology* provides a variety of worksheets that are correlated directly with each chapter of the program. Representations of these worksheets are found at the beginning of each chapter in this Annotated Teacher's Edition. Specific worksheets related to each chapter are listed in the Chapter Organizer. Worksheets and transparencies are found in the softcover *Teaching Resources* for each course.

Many worksheets are also available on the HRW Web site. The address is **go.hrw.com**.

Science Skills Worksheets: Thinking Skills

BEING FLEXIBLE
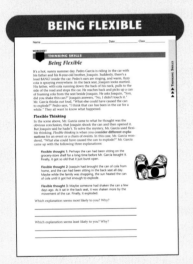

USING YOUR SENSES
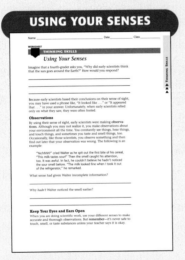

THINKING OBJECTIVELY
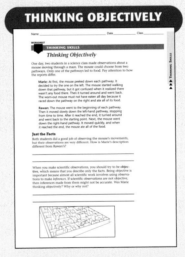

UNDERSTANDING BIAS
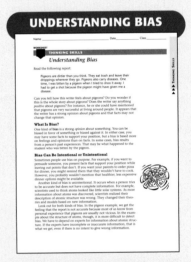

USING LOGIC

BOOSTING YOUR MEMORY
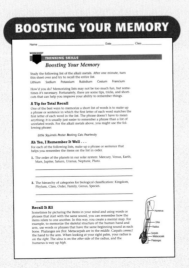

IMPROVING YOUR STUDY HABITS
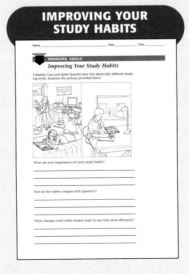

READING A SCIENCE TEXTBOOK
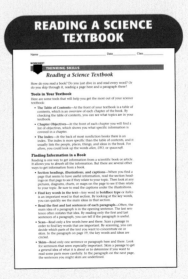

Science Skills Worksheets: Experimenting Skills

SAFETY RULES!

DOING A LAB WRITE-UP
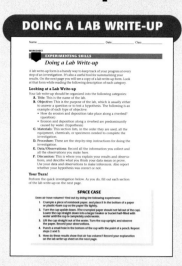

UNDERSTANDING VARIABLES
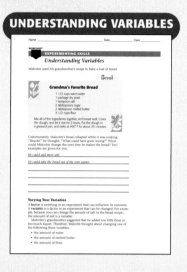

WORKING WITH HYPOTHESES
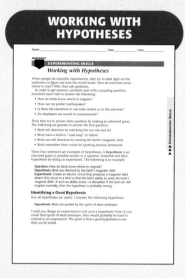

DESIGNING AN EXPERIMENT
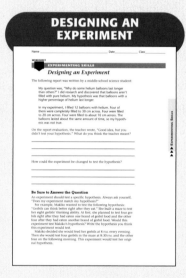

USING THE INTERNATIONAL SYSTEM OF UNITS (SI)
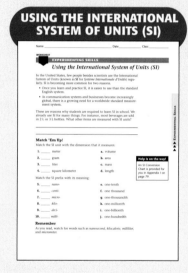

MEASURING
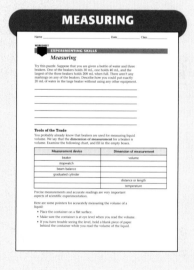

Science Skills Worksheets: Researching Skills

CHOOSING YOUR TOPIC
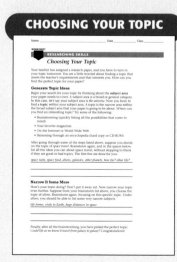

ORGANIZING YOUR RESEARCH
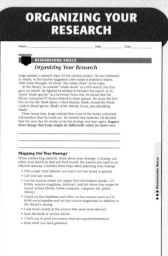

FINDING USEFUL SOURCES

RESEARCHING ON THE WEB
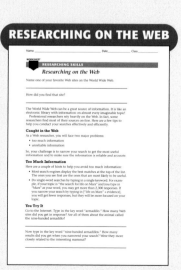

Science & Math Skills Worksheets (continued)

Science Skills Worksheets: Researching Skills (continued)

IDENTIFYING BIAS

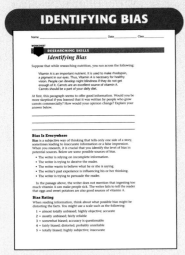

TAKING NOTES

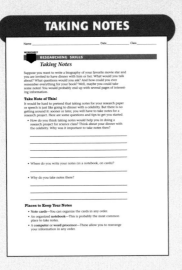

SCIENCE WRITING

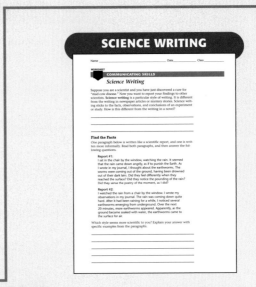

Science Skills Worksheets: Communicating Skills

SCIENCE DRAWING

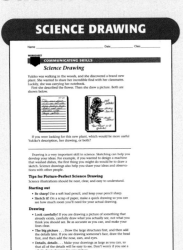

USING MODELS TO COMMUNICATE

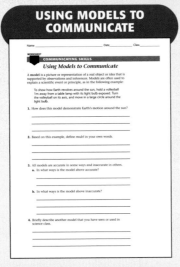

INTRODUCTION TO GRAPHS

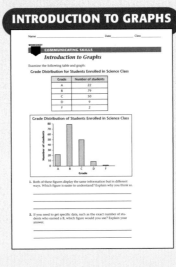

GRASPING GRAPHING

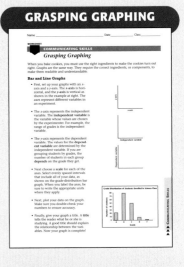

INTERPRETING YOUR DATA

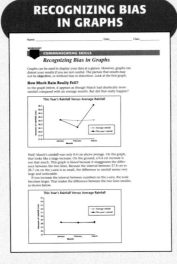

RECOGNIZING BIAS IN GRAPHS

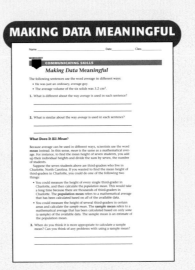

MAKING DATA MEANINGFUL

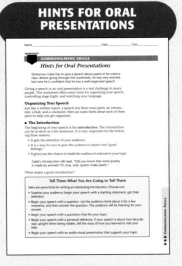

HINTS FOR ORAL PRESENTATIONS

Math Skills for Science

ADDITION AND SUBTRACTION

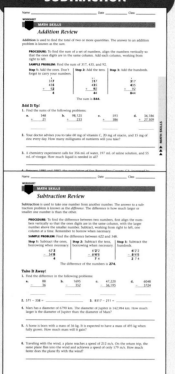

WORKSHEET — MATH SKILLS
Addition Review

Name _____ Date _____ Class _____

Addition is used to find the total of two or more quantities. The answer to an addition problem is known as the *sum*.

PROCEDURE: To find the sum of a set of numbers, align the numbers vertically so that the ones digits are in the same column. Add each column, working from right to left.

SAMPLE PROBLEM: Find the sum of 317, 435, and 92.

Step 1: Add the ones. Don't forget to carry your numbers.
Step 2: Add the tens.
Step 3: Add the hundreds.

```
    4            1
  317          317          317
  435          435          435
 + 92         + 92         + 92
   4            44           844
```
The sum is **844**.

Add It Up!
1. Find the sums of the following problems.
   a. 348 21   b. 98,125 233   c. 593 386   d. 36,186 27,309

2. Your doctor advises you to take 60 mg of vitamin C, 20 mg of niacin, and 15 mg of zinc every day. How many milligrams of nutrients will you take?

3. A chemistry experiment calls for 356 mL of water, 197 mL of saline solution, and 55 mL of vinegar. How much liquid is needed in all?

4. Between 1980 and 1985, the population of San Bernardino County, CA, increased by...

WORKSHEET — MATH SKILLS
Subtraction Review

Subtraction is used to take one number from another number. The answer to a subtraction problem is known as the *difference*. The difference must be smaller than the larger number or smaller than the one you start with.

PROCEDURE: To find the difference between two numbers, first align the numbers vertically so that the ones digits are in the same column, with the larger number above the smaller number. Subtract, working from right to left, one column at a time. Remember to borrow when necessary.

SAMPLE PROBLEM: Find the difference between 622 and 348.

Step 1: Subtract the ones, borrowing when necessary.
Step 2: Subtract the tens, borrowing when necessary.
Step 3: Subtract the hundreds.

The difference of the numbers is **274.**

Take It Away!
1. Find the difference in the following problems.
   a. 88 36   b. 1695 352   c. 846 36,195   d. 6048 3724

2. 571 − 338 =   3. 8317 − 211 =

4. Mars has a diameter of 6790 km. The diameter of Jupiter is 142,984 km. How much larger is the diameter of Jupiter than the diameter of Mars?

5. A horse is born with a mass of 36 kg. It is expected to have a mass of 495 kg when fully grown. How much mass will it gain?

6. Traveling with the wind, a plane reaches a speed of 212 m/s. On the return trip, the same plane flies into the wind and achieves a speed of only 179 m/s. How much faster does the plane fly with the wind?

MULTIPLICATION

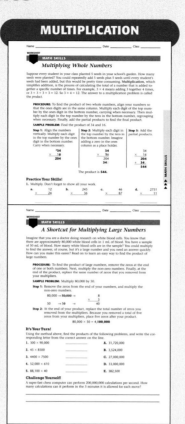

WORKSHEET — MATH SKILLS
Multiplying Whole Numbers

Suppose every student in your class planted 5 seeds in your school's garden. How many seeds were planted? You could repeatedly add 5 seeds plus 5 seeds until every student's seeds had been added, but this would be pretty time consuming. **Multiplication**, which simplifies addition, is the process of calculating the total of a number that is added together a specific number of times. For example, 5 × 3 = 15 or 3 × 5 = 15. So too is adding 5 together 3 times, 5 + 5 + 5 = 15. The answer to a multiplication problem is called the *product*.

PROCEDURE: To find the product of two whole numbers, align your numbers so that the ones digits are in the same column. Multiply each digit of the top number by the ones digit in the bottom number, carrying when necessary. Then multiply each digit in the top number by the tens in the bottom number, regrouping when necessary. Finally, add the partial products to find the final product.

SAMPLE PROBLEM: Find the product of 34 and 16.

Step 1: Align the numbers vertically. Multiply each digit in the top number by the ones digit in the bottom number. Carry when necessary.
Step 2: Multiply each digit in the top number by the tens in the bottom number. Imagine adding a zero in the ones column as a place holder.
Step 3: Add the partial products.

The product is **544.**

Practice Your Skills!
1. Multiply. Don't forget to show all your work.
   a. 24   b. 36   c. 46 87   d. 2751 11

WORKSHEET — MATH SKILLS
A Shortcut for Multiplying Large Numbers

Imagine that you are a doctor doing research on white blood cells. You know that there are approximately 80,000 white blood cells in 1 mL of blood. You have a sample of 50 mL of blood. How many white blood cells are in the sample? You could multiply to find the answer, of course, but it's a large number and you need an answer quickly. How can you make this easier? Read on to learn an easy way to find the product of large numbers.

PROCEDURE: To find the product of large numbers, remove the zeros at the end of one or both numbers. Next, multiply the non-zero numbers. Finally, at the end of the product, replace the same number of zeros that you removed from your multipliers.

SAMPLE PROBLEM: Multiply 80,000 by 50.

Step 1: Remove the zeros from the end of your numbers, and multiply the non-zero numbers.
80,000 → 8   50 → 5
8 × 5 = 40

Step 2: At the end of your product, replace the total number of zeros you removed from the multipliers. Because you removed a total of five zeros from your multipliers, place five zeros after your product.
80,000 × 50 = 4,000,000

It's Your Turn!
Using the method above, find the products of the following problems, and write the corresponding letter from the correct answer on the line.
1. 100 × 90,000 _____   A. 31,720,000
2. 45 × 8500 _____   B. 3,524,000
3. 4400 × 7500 _____   C. 27,000,000
4. 52,000 × 610 _____   D. 33,000,000
5. 88,100 × 40 _____   E. 382,500

Challenge Yourself!
A super-fast chess computer can perform 200,000,000 calculations per second. How many calculations can it perform in the 3 minutes it is allowed for each move?

DIVISION

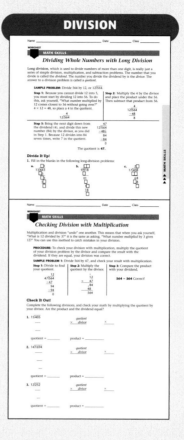

WORKSHEET — MATH SKILLS
Dividing Whole Numbers with Long Division

Long division, which is used to divide numbers of more than one digit, is really just a series of simple division, multiplication, and subtraction problems. The number that you divide is called the *dividend.* The number you divide the dividend by is the *divisor.* The answer to a division problem is called a *quotient.*

SAMPLE PROBLEM: Divide 564 by 12, or 12)564

Step 1: Because you cannot divide 12 into 5, you must start by dividing 12 into 56. To do this, ask yourself, "What number multiplied by 12 comes closest to 56 without going over?" 4 × 12 = 48, so place a 4 in the quotient.

Step 2: Multiply the 4 by the divisor and place the product under the 56. Then subtract that product from 56.

Step 3: Bring the next digit down from the dividend (4), and divide that new number (84) by the divisor, as you did in Step 1. Because 12 divides into 84 seven times, write 7 in the quotient.

The quotient is **47.**

Divide It Up!
1. Fill in the blanks in the following long-division problems.
   a. 13)563   b. 9)918   c. 17)408

WORKSHEET — MATH SKILLS
Checking Division with Multiplication

Multiplication and division "undo" one another. This means that when you ask yourself, "What is 12 divided by 37?" it is the same as, "What number *multiplied* by 3 gives 12?" You can use this method to catch mistakes in your division.

PROCEDURE: To check your division with multiplication, multiply the quotient of your division problem by the divisor and compare the result with the dividend. If they are equal, your division was correct.

SAMPLE PROBLEM: Divide 564 by 47, and check your result with multiplication.

Step 1: Divide to find your quotient.
Step 2: Multiply the quotient by the divisor.
Step 3: Compare the product with your dividend.

564 = 564 Correct!

Check It Out!
Complete the following divisions, and check your math by multiplying the quotient by your divisor. Are the product and the dividend equal?

1. 15)405   quotient _____ × divisor _____
   quotient _____ = product _____

2. 14)694   quotient _____ × divisor _____
   quotient _____ = product _____

3. 12)252   quotient _____ × divisor _____
   quotient _____ = product _____

AVERAGES

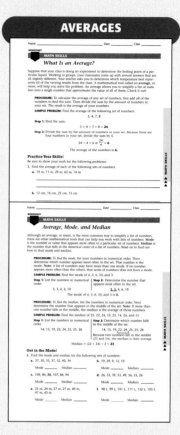

WORKSHEET — MATH SKILLS
What Is an Average?

Suppose that your class is doing an experiment to determine the boiling point of a particular liquid. Working in groups, your classmates come up with several answers that are all slightly different. Your teacher asks you to determine which temperature best represents all of the varying results from the class. A mathematical tool called an *average,* or *mean,* will help you solve the problem. An average allows you to simplify a list of numbers into a single number that *approximates* the value of all of them. Check it out!

PROCEDURE: To calculate the average of any set of numbers, first add all of the numbers to find the sum. Then divide the sum by the amount of numbers in your set. The result is the average of your numbers.

SAMPLE PROBLEM: Find the average of the following set of numbers:
5, 4, 7, 8

Step 1: Find the sum.
5 + 4 + 7 + 8 = 24

Step 2: Divide the sum by the amount of numbers in your set. Because there are four numbers in your set, divide the sum by 4.
24 ÷ 4 = 6 or 24/4 = 6
The average of the numbers is **6.**

Practice Your Skills!
Be sure to show your work for the following problems.
1. Find the average of the following sets of numbers.
   a. 19 m, 11 m, 29 m, 62 m, 14 m
   b. 12 cm, 16 cm, 25 cm, 15 cm

WORKSHEET — MATH SKILLS
Average, Mode, and Median

Although an average, or mean, is the most common way to simplify a list of numbers, there are other mathematical tools that can help you work with lists of numbers. **Mode** is the number or value that appears most often in a particular set of numbers. **Median** is the number that falls in the *numerical center* of a list of numbers. Read on to find out how to find mode and median.

SAMPLE PROBLEM: Find the mode of 4, 3, 6, 10, and 3.

Step 1: List the numbers in numerical order.
3, 3, 4, 6, 10
Step 2: Determine which number appears most often in the set.
3, 3, 4, 6, 10
The mode of 4, 3, 6, 10, and 3 is **3.**

PROCEDURE: To *find the median,* list the numbers in numerical order. Next determine the number that appears in the middle of the set. Note: If more than one number falls in the middle, the median is the average of those numbers.

SAMPLE PROBLEM: Find the median of 25, 22, 24, 19, 25, 14, 26, and 15.

Get in the Mode!
1. Find the mode and median for the following sets of numbers.
   a. 37, 30, 35, 37, 32, 40, 34   Mode _____ Median _____
   b. 48, 39, 22, 94, 9, 12, 10   Mode _____ Median _____
   c. 109, 84, 88, 107, 84, 94   Mode _____ Median _____
   d. 26, 53, 39, 53, 49, 56, 15, 26 Mode _____ Median _____
   e. 25 m, 24 m, 27 m, 27 m, 49 m, 47 m, 45 m   Mode _____ Median _____
   f. 98 L, 99 L, 101 L, 111 L, 132 L, 103 L   Mode _____ Median _____

POSITIVE AND NEGATIVE NUMBERS

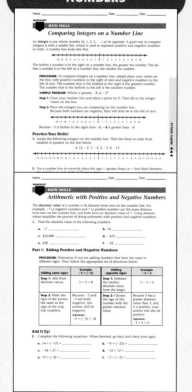

WORKSHEET — MATH SKILLS
Comparing Integers on a Number Line

An **integer** is any whole number (0, 1, 2, 3, . . .) or its opposite. A good way to compare integers is with a *number line,* which is used to represent positive and negative numbers in order. A number line looks like this:

The farther a number is to the right on a number line, the greater the number. The farther a number is to the left on a number line, the smaller the number.

PROCEDURE: To compare integers on a number line, simply place your values on the line, with positive numbers to the right of zero and negative numbers to the left of zero. The number that is the farthest to the right is the greatest number. The number that is the farthest to the left is the smallest number.

SAMPLE PROBLEM: Which is greater, −8 or −3?

Step 1: Draw your number line and select a point for 0. Then fill in the integer values on the line.
Step 2: Place the integers you are comparing on the number line. Because both numbers are negative, they will both be to the left of zero.

Because −3 is farther to the right than −8, **−3** is greater than −8.

Practice Your Skills!
1. Locate the following integers on the number line. Then list them in order from smallest to greatest on the line below.
4, 12, −2, 7, −5, 2, −7, 9, −13

2. Use a number line to correctly place the sign > (greater than) or < (less than) between...

WORKSHEET — MATH SKILLS
Arithmetic with Positive and Negative Numbers

The **absolute value** of a number is its distance from zero on the number line. For example, −7 (a negative number) and 7 (a positive number) are the same distance from zero on the number line, and both have an absolute value of 7. Using absolute values simplifies the process of doing arithmetic with positive and negative numbers.
1. Find the absolute value of the following numbers.
   a. −7 _____   b. 14 _____
   c. 325,000 _____   d. −1 _____
   e. 230 _____   f. −52 _____

Part 1: Adding Positive and Negative Numbers
PROCEDURE: Determine if you are adding numbers that have the same or different signs. Then follow the appropriate set of directions below.

	Example −3 + (−5)	Adding opposite signs	Example −3 + 5
Adding same signs			
Step 1: Add their absolute values.	3 + 5 = 8	**Step 1:** Subtract the smaller absolute value from the larger.	5 − 3 = 2
Step 2: Make the sign of the answer the same as the sign of the original numbers.	Because −3 and −5 are both negative, the answer will be negative. Answer: −3 + (−5) = −8	**Step 2:** Choose the sign of the number with the greater absolute value.	Because 5 has a greater absolute value than 3 and 5 is positive, your answer will also be positive. Answer: −3 + 5 = 2

Add It Up!
2. Complete the following equations. When finished, go back and check your signs.
   a. 14 + (−17) =   b. 30 + (−23) =
   c. −16 + 21 =   d. −12 + 12 =
   e. 15 + (−4) =   f. −7 + (−7) =

FRACTIONS

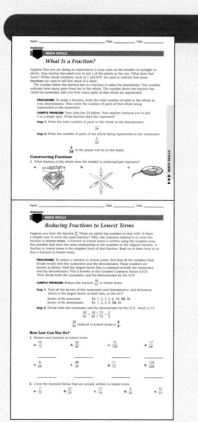

WORKSHEET — MATH SKILLS
What Is a Fraction?

Suppose that you are doing an experiment in your class on the benefits of sunlight to plants. Your teacher has asked you to put ⅔ of the plants in the sun. What does that mean? While whole numbers, such as 1 and 879, are used to indicate *how many,* **fractions** are used to tell *how much of a whole.*

The number below the fraction bar in a fraction is called the *denominator.* This number indicates how many equal parts make up the whole. The number above the fraction bar, called the *numerator,* tells you how many parts of that whole are represented.

PROCEDURE: To make a fraction, write the total number of units in the whole as the denominator. Then write the number of parts of that whole being represented as the numerator.

SAMPLE PROBLEM: Your class has 24 plants. Your teacher instructs you to put 5 in a shady spot. What fraction does this represent?

Step 1: Write the total number of parts in the whole as the denominator.
_/24

Step 2: Write the number of parts of the whole being represented as the numerator.
5/24

5/24 of the plants will be in the shade.

Constructing Fractions
1. What fraction of the whole does the shaded or patterned part represent?
a. _____   b. _____   c. _____

WORKSHEET — MATH SKILLS
Reducing Fractions to Lowest Terms

Suppose you have the fraction 10/30. Those are pretty big numbers to deal with. Is there a simpler way to write the same fraction? Well, one common method is to write the fraction in **lowest terms.** A fraction in lowest terms is written using the smallest numbers possible that have the same relationship as the numbers of the original fraction. A fraction in lowest terms is the simplest form of that fraction. Read on to learn how to reduce a fraction to lowest terms.

PROCEDURE: To reduce a fraction to lowest terms, first find all the numbers that divide evenly into the numerator and denominator, called *factors.* Find the largest factor that is common to both the numerator and the denominator. This is known as the Greatest Common Factor (GCF). Then divide both the numerator and the denominator by the GCF.

SAMPLE PROBLEM: Reduce the fraction 10/30 to lowest terms.

Step 1: Find all the factors of the numerator and denominator, and determine which is the largest factor in both lists, or the GCF.
factors of the numerator 30: 1, 2, 3, 5, 6, 10, **15**, 30
factors of the denominator 45: 1, 3, 5, 9, **15**, 45

Step 2: Divide both the numerator and the denominator by the GCF, which is 15.
30/45 = 30 ÷ 15 / 45 ÷ 15
reduced to lowest terms is **2/3.**

How Low Can You Go?
1. Reduce each fraction to lowest terms.
   a. 10/12   b. 36/60   c. 75/100   d. 17/68
   e. 8/64   f. 48/56   g. 11/15   h. 150/200

2. Circle the fractions below that are already written in lowest terms.
   a. 7/77   b. 21/25   c. 9/20   d. 37/51

WORKSHEET — MATH SKILLS
Improper Fractions and Mixed Numbers

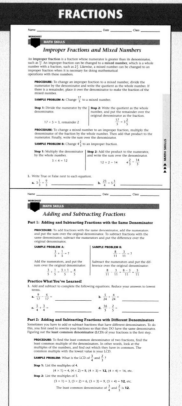

An **improper fraction** is a fraction whose numerator is greater than its denominator, such as 17/5. An improper fraction can be changed to a **mixed number,** which is a whole number with a fraction, such as 3 2/5. Likewise, a mixed number can be changed to an improper fraction when it is necessary for doing mathematical operations with these numbers.

PROCEDURE: To change an improper fraction to a mixed number, divide the numerator by the denominator and write the quotient as the whole number. If there is a remainder, place it over the denominator to make the fraction of the mixed number.

SAMPLE PROBLEM: Change 17/5 to a mixed number.

Step 1: Divide the numerator by the denominator.
17 ÷ 5 = 3, remainder 2
Step 2: Write the quotient as the whole number, and the remainder over the original denominator as the fraction.
17/5 = 3 2/5

PROCEDURE: To change a mixed number to an improper fraction, multiply the denominator of the fraction by the whole number. Then add that product to the numerator. Finally, write the sum over the denominator.

SAMPLE PROBLEM: Change 4 2/3 to an improper fraction.

Step 1: Multiply the denominator by the whole number.
3 × 4 = 12
Step 2: Add the product to the numerator, and write the sum over the denominator.
12 + 2 = 14
4 2/3 = 14/3

1. Write True or False next to each equation.
   a. 3 1/3 = 5/3   b. 27/4 = 6 3/4

WORKSHEET — MATH SKILLS
Adding and Subtracting Fractions

Part 1: Adding and Subtracting Fractions with the Same Denominator
PROCEDURE: To add fractions with the same denominator, add the numerators and put the sum over the original denominator. To subtract fractions with the same denominator, subtract the numerators and put the difference over the original denominator.

SAMPLE PROBLEM A:
3/5 + 1/5
Add the numerators, and put the sum over the original denominator:
3/5 + 1/5 = 4/5

SAMPLE PROBLEM B:
8/9 − 3/9
Subtract the numerators and put the difference over the original denominator:
8/9 − 3/9 = 5/9

Practice What You've Learned!
1. Add and subtract to complete the following equations. Reduce your answers to lowest terms.
   a. 9/17 − 6/17 =   b. 21/24 − 3/24 =
   c. 3/4 − 1/4 =   d. 16/5 − 2/5 =

Part 2: Adding and Subtracting Fractions with Different Denominators
Sometimes you have to add or subtract fractions that have different denominators. To do this, you first need to rewrite your fractions so that they DO have the same denominator. Figuring out the **least common denominator (LCD)** of your fractions is the first step.

PROCEDURE: To find the least common denominator of two fractions, find the least common multiple of the denominators. In other words, look at the multiples of the numbers, and find out which they have in common. The common multiple with the lowest value is your LCD.

SAMPLE PROBLEM: What is the LCD of 3/4 and 5/3?

Step 1: List the multiples of 4.
(4 × 1) = 4, (4 × 2) = 8, (4 × 3) = **12**, (4 × 4) = 16, etc.
Step 2: List the multiples of 3.
(3 × 1) = 3, (3 × 2) = 6, (3 × 3) = 9, (3 × 4) = **12**, etc.
The least common denominator of 3/4 and 5/3 is **12.**

WORKSHEET — MATH SKILLS
Multiplying and Dividing Fractions

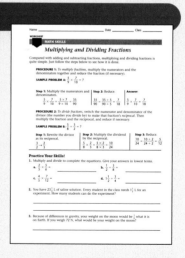

Compared with adding and subtracting fractions, multiplying and dividing fractions is quite simple. Just follow the steps below to see how it is done.

PROCEDURE: To *multiply fractions,* multiply the numerators and the denominators together and reduce the fraction (if necessary).

SAMPLE PROBLEM A: 5/9 × 7/2 = ?

Step 1: Multiply the numerators and denominators.
5/9 × 7/2 = 5 × 7 / 9 × 2 = 35/18
Step 2: Reduce.
35/18 = 35 ÷ 5 / 9 × 2 = 7/18
Answer:
5/9 × 7/2 = 35/18

PROCEDURE: To *divide fractions,* multiply the numerator and denominator of the divisor (the second number you divide by) to make that fraction's reciprocal. Then multiply the first number and the reciprocal, and reduce if necessary.

SAMPLE PROBLEM: 8/3 ÷ 2/1 = ?

Step 1: Rewrite the divisor as its reciprocal.
3/2
Step 2: Multiply the dividend by the reciprocal.
8/3 × 5/2 × 10/8 = ...
Step 3: Reduce.
10/24 = 10 ÷ 2 / 24 ÷ 2 = 5/12

Practice Your Skills!
1. Multiply and divide to complete the equations. Give your answers in lowest terms.
   a. 2/5 × 5/9 =   b. 1/2 ÷ 4/5 =
   c. 4/9 × 7/12 =   d. 1 3/4 × 1/4 =

2. You have 23¼ L of saline solution. Every student in the class needs 1¼ L for an experiment. How many students can do the experiment?

3. Because of differences in gravity, your weight on the moon would be ⅙ what it is on Earth. If you weigh 72 N, what would be your weight on the moon?

Science & Math Skills Worksheets (continued)

Math Skills for Science (continued)

RATIOS AND PROPORTIONS

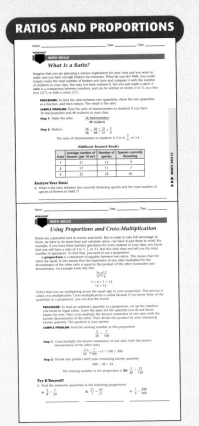

DECIMALS

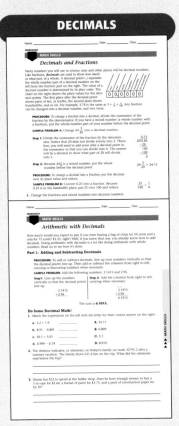

PERCENTAGES

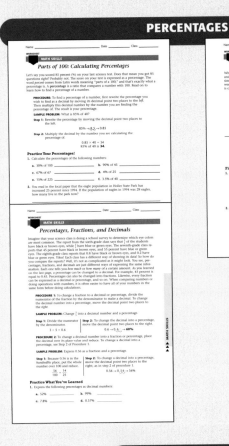

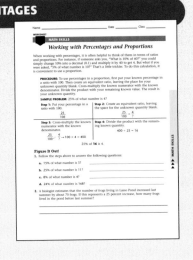

POWERS OF 10

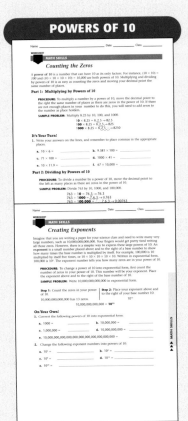

SCIENTIFIC NOTATION

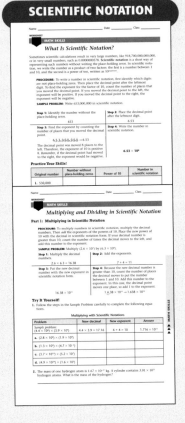

SI MEASUREMENT AND CONVERSION

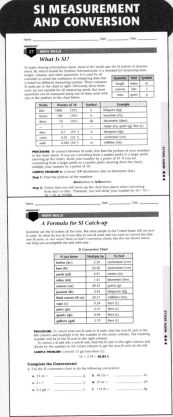

Math Skills for Science (continued)

GEOMETRY

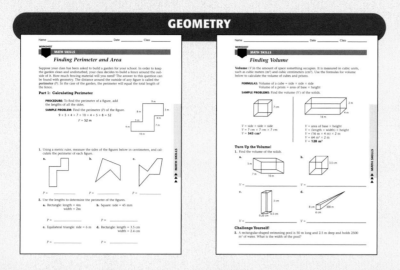

THE UNIT FACTOR AND DIMENSIONAL ANALYSIS

MATH IN SCIENCE: INTEGRATED SCIENCE

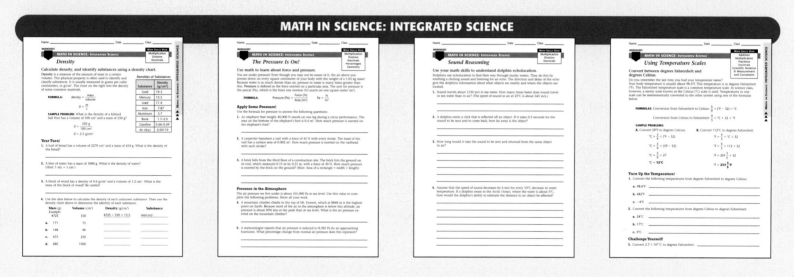

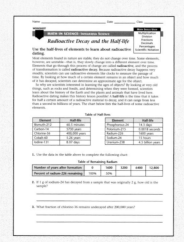

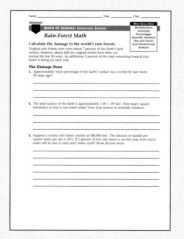

Science & Math Skills Worksheets (continued)

Math Skills for Science (continued)

MATH IN SCIENCE: EARTH SCIENCE

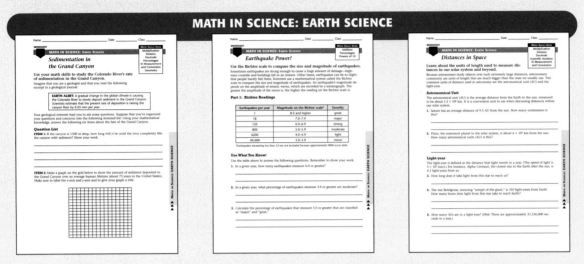

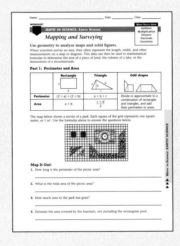

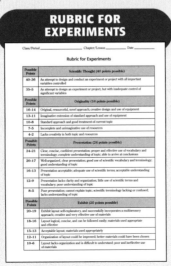

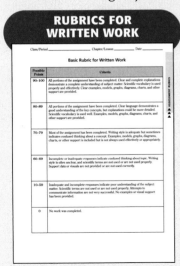

Assessment Checklist & Rubrics

The following is just a sample of over 50 checklists and rubrics contained in this booklet.

EARTH SCIENCE NATIONAL SCIENCE EDUCATION STANDARDS CORRELATIONS

The following lists show the chapter correlation of **Holt Science and Technology: Weather and Climate** with the *National Science Education Standards* (grades 5-8)

UNIFYING CONCEPTS AND PROCESSES

Standard	Chapter Correlation	
Evidence, models, and explanation Code: UCP 2	Chapter 1 Chapter 2 Chapter 3	1.2 2.1 3.2, 3.3
Change, constancy, and measurement Code: UCP 3	Chapter 1 Chapter 2 Chapter 3	1.3 2.1, 2.4 3.2, 3.3

SCIENCE AS INQUIRY

Standard	Chapter Correlation	
Abilities necessary to do scientific inquiry Code: SAI 1	Chapter 1 Chapter 2 Chapter 3	1.1, 1.2, 1.3, 1.4 2.1, 2.4 3.1, 3.2, 3.3
Understandings about science and technology Code: SAI 2	Chapter 1 Chapter 3	1.2 3.3

HISTORY AND NATURE OF SCIENCE

Standard	Chapter Correlation	
Science as a human endeavor Code: HNS 1	Chapter 2 Chapter 3	2.4 3.4, 3.3
History of science Code: HNS 3	Chapter 2 Chapter 3	2.2 3.2

SCIENCE IN PERSONAL AND SOCIAL PERSPECTIVES

Standard	Chapter Correlation	
Personal health Code: SPSP 1	Chapter 1 Chapter 2	1.1, 1.4 2.1
Populations, resources, and environments Code: SPSP 2	Chapter 1	1.2, 1.4
Natural hazards Code: SPSP 3	Chapter 1 Chapter 2 Chapter 3	1.1, 1.2, 1.4 2.1, 2.3 3.1, 3.3
Risks and benefits Code: SPSP 4	Chapter 1 Chapter 2 Chapter 3	1.1, 1.2, 1.4 2.3 3.3
Science and technology in society Code: SPSP 5	Chapter 1 Chapter 2 Chapter 3	1.4 2.4 3.3

SCIENCE AND TECHNOLOGY

Standard	Chapter Correlation	
Abilities of technological design Code: ST 1	Chapter 1 Chapter 2 Chapter 3	1.1, 1.2, 1.3 2.4 3.2
Understandings about science and technology Code: ST 2	Chapter 2 Chapter 3	2.4 3.3

EARTH SCIENCE National Science Education Content Standards

STRUCTURE OF THE EARTH SYSTEM

Standard	Chapter Correlation	
Water, which covers the majority of the earth's surface, circulates through the crust, oceans, and atmosphere in what is known as the "water cycle." Water evaporates from the earth's surface, rises and cools as it moves to higher elevations, condenses as rain or snow, and falls to the surface where it collects in lakes, oceans, soil, and in rocks underground. Code: ES 1f	**Chapter 2**	2.1
The atmosphere is a mixture of nitrogen, oxygen, and trace gases that include water vapor. The atmosphere has different properties at different elevations. Code: ES 1h	**Chapter 1**	1.1
Clouds, formed by the condensation of water vapor, affect weather and climate. Code: ES 1i	**Chapter 2**	2.1, 2.3
Global patterns of atmospheric movement influence local weather. Oceans have a major effect on climate, because water in the oceans holds a large amount of heat. Code: ES 1j	**Chapter 1** **Chapter 2** **Chapter 3**	1.3 2.2, 2.3 3.1
Living organisms have played many roles in the earth system, including affecting the composition of the atmosphere, producing some types of rocks, and contributing to the weathering of rocks. Code: ES 1k	**Chapter 1** **Chapter 3**	1.2, 1.4 3.3

EARTH'S HISTORY

Standard	Chapter Correlation	
The earth processes we see today, including erosion, movement of lithospheric plates, and changes in atmospheric composition, are similar to those that occurred in the past. Earth history is also influenced by occasional catastrophes, such as the impact of an asteroid or comet. Code: ES 2a	**Chapter 1** **Chapter 3**	1.2, 1.4 3.3

EARTH IN THE SOLAR SYSTEM

Standard	Chapter Correlation	
The sun is the major source of energy for phenomena on the earth's surface, such as growth of plants, winds, ocean currents, and the water cycle. Seasons result from variations in the amount of the sun's energy hitting the surface, due to the tilt of the earth's rotation on its axis and the length of the day. Code: ES 3d	**Chapter 1** **Chapter 3**	1.3 3.1

Master Materials List

For added convenience, Science Kit® provides materials-ordering software on CD-ROM designed specifically for *Holt Science and Technology*. Using this software, you can order complete kits or individual items, quickly and efficiently.

CONSUMABLE MATERIALS	AMOUNT	PAGE
Balloon	1	26, 100
Card, index, 3 x 5 in.	1	26, 58, 100
Cardboard, corrugated, 15 x 15 cm	4	109
Cup, paper	5	102
Cup, plastic-foam, large	2	100
Fan, paper	1	71
Filters, coffee, cone-shaped (or plastic funnel)	1	100
Food coloring, red	1 bottle	16, 100
Ice, cubed	2	39, 71, 100
Ice, shaved	150 mL	107
Marker, colored	1	102
Oil, cooking	500 mL	35
Paint, black tempera	1 container	109
Paint, light blue tempera	1 container	109
Paint, white tempera	1 container	109
Pencil, colored	4	108
Plastic tubing, 5 mm diam., 30 cm long (or clear inflexible plastic straws)	1	100
Plate, paper	1	58
Rubber band	1	26
Straw	1	26
Straw, straight plastic	1	58
Straw, straight plastic	2	102
String	1 ball	16
Tape, masking	1 roll	26, 102
Tape, transparent	1 roll	26, 100

NONCONSUMABLE EQUIPMENT	AMOUNT	PAGE
Air Pump	1	3
Balance, triple-beam, metric (or scale)	1	3
Basketball, or football, deflated	1	3
Beaker, 100 mL	1	107
Beaker, 1000 mL	2	35
Bottle, small (vitamin)	1	16, 100
Can, coffee, 10 cm diam.	1	26
Can, soda	1	100
Clay, modeling	1 lb	100, 102
Compass, drawing	1	58
Compass, magnetic	1	58
Container, plastic	1	16
Container, plastic (or jar or glass)	1	39
Container, plastic, rectangular	1	35
Cup, plastic	1	71
Film canister	1	100
Globe, world	1	67
Gloves, heat-resistant	1 pair	100, 107
Graduated cylinder, 100 mL	1	107
Hole punch	1	102
Hot plate	1	100, 107
Lamp, goose-neck, 60 W bulb	1	67
Pan, aluminum pie	1	100
Pitcher	1	100
Protractor	1	58
Putty, adhesive	1 package	67
Rock, small, approx. 1/4 lb	1	58
Rubber square, tan, approx. 15 x 15 cm	1	109
Ruler, metric	1	58, 100, 102, 107, 108
Scissors	1	26, 58, 102
Stapler	1	58
Stapler (mini)	1	102
Stopwatch	1	102, 109
Test tube	1	100
Thermometer, Celsius	1	71, 100
Thermometer, Celsius	2	67
Thermometer, Celsius	4	109
Thumbtack	1	102
Thumbtack (or pushpin)	1	58
Wood square, approx. 15 x 15 cm (beige or tan)	1	109
Yogurt container with lid	1	100

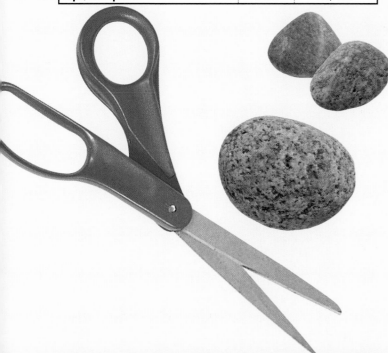

Answers to Concept Mapping Questions

The following pages contain sample answers to all of the concept mapping questions that appear in the Chapter Reviews. Because there is more than one way to do a concept map, your students' answers may vary.

CHAPTER 1 The Atmosphere

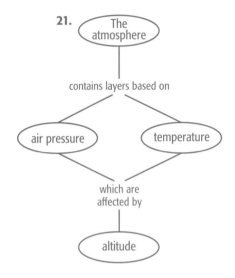

21.
The atmosphere

contains layers based on

air pressure — temperature

which are affected by

altitude

CHAPTER 2 Understanding Weather

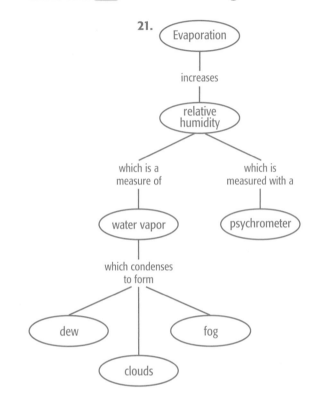

21.
Evaporation

increases

relative humidity

which is a measure of — which is measured with a

water vapor — psychrometer

which condenses to form

dew — fog

clouds

CHAPTER 3 Climate

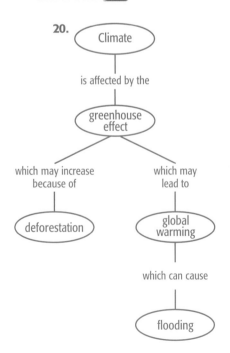

20.
Climate

is affected by the

greenhouse effect

which may increase because of — which may lead to

deforestation — global warming

which can cause

flooding

To the Student

This book was created to make your science experience interesting, exciting, and fun!

Go for It!

Science is a process of discovery, a trek into the unknown. The skills you develop using *Holt Science & Technology*— such as observing, experimenting, and explaining observations and ideas— are the skills you will need for the future. There is a universe of exploration and discovery awaiting those who accept the challenges of science.

Science & Technology

You see the interaction between science and technology every day. Science makes technology possible. On the other hand, some of the products of technology, such as computers, are used to make further scientific discoveries. In fact, much of the scientific work that is done today has become so technically complicated and expensive that no one person can do it entirely alone. But make no mistake, the creative ideas for even the most highly technical and expensive scientific work still come from individuals.

Activities and Labs

The activities and labs in this book will allow you to make some basic but important scientific discoveries on your own. You can even do some exploring on your own at home! Here's your chance to use your imagination and curiosity as you investigate your world.

Keep a ScienceLog

In this book, you will be asked to keep a type of journal called a ScienceLog to record your thoughts, observations, experiments, and conclusions. As you develop your ScienceLog, you will see your own ideas taking shape over time. You'll have a written record of how your ideas have changed as you learn about and explore interesting topics in science.

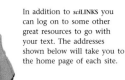

Know "What You'll Do"

The "What You'll Do" list at the beginning of each section is your built-in guide to what you need to learn in each chapter. When you can answer the questions in the Section Review and Chapter Review, you know you are ready for a test.

Check Out the Internet

You will see this logo throughout the book. You'll be using *sci*LINKS as your gateway to the Internet. Once you log on to *sci*LINKS using your computer's Internet link, type in the *sci*LINKS address. When asked for the keyword code, type in the keyword for that topic. A wealth of resources is now at your disposal to help you learn more about that topic.

In addition to *sci*LINKS you can log on to some other great resources to go with your text. The addresses shown below will take you to the home page of each site.

internet**connect**

This textbook contains the following on-line resources to help you make the most of your science experience.

go.hrw.com	***sci*LINKS** NSTA	**Smithsonian Institution®** Internet Connections	**CNNfyi.com**
Visit **go.hrw.com** for extra help and study aids matched to your textbook. Just type in the keyword HR2 HOME.	Visit **www.scilinks.org** to find resources specific to topics in your textbook. Keywords appear throughout your book to take you further.	Visit **www.si.edu/hrw** for specifically chosen on-line materials from one of our nation's premier science museums.	Visit **www.cnnfyi.com** for late-breaking news and current events stories selected just for you.

Chapter Organizer

CHAPTER ORGANIZATION	TIME MINUTES	OBJECTIVES	LABS, INVESTIGATIONS, AND DEMONSTRATIONS
Chapter Opener pp. 2–3	45	National Standards: SAI 1	**Start-Up Activity,** Air—It's Massive, p. 3
Section 1 Characteristics of the Atmosphere	90	▶ Discuss the composition of the Earth's atmosphere. ▶ Explain why pressure changes with altitude. ▶ Explain how temperature changes with altitude. ▶ Describe the layers of the atmosphere. SPSP 1, 3, 4, ES 1h; Labs SAI 1, ST 1	**Demonstration,** Air Pressure, p. 4 in ATE **Discovery Lab,** Under Pressure! p. 26 **Datasheets for LabBook,** Under Pressure! **Whiz-Bang Demonstrations,** Blue Sky
Section 2 Heating of the Atmosphere	90	▶ Describe what happens to radiation that reaches the Earth. ▶ Summarize the processes of radiation, conduction, and convection. ▶ Explain how the greenhouse effect could contribute to global warming. UCP 2, SAI 2, SPSP 2–4, ES 1k, 2a; Labs SAI 1, ST 1	**Demonstration,** p. 11 in ATE **Design Your Own,** Boiling Over! p. 100 **Datasheets for LabBook,** Boiling Over! **Calculator-Based Labs,** The Greenhouse Effect **EcoLabs & Field Activities,** That Greenhouse Effect!
Section 3 Atmospheric Pressure and Winds	90	▶ Explain the relationship between air pressure and wind direction. ▶ Describe the global patterns of wind. ▶ Explain the causes of local wind patterns. SAI 1, ES 1j, 3d; Labs UCP 3, SAI 1, ST 1	**Demonstration,** Air Movement, p. 14 in ATE **QuickLab,** Full of "Hot Air," p. 16 **Discovery Lab,** Go Fly a Bike! p. 102 **Datasheets for LabBook,** Go Fly a Bike!
Section 4 The Air We Breathe	90	▶ Describe the major types of air pollution. ▶ Name the major causes of air pollution. ▶ Explain how air pollution can affect human health. ▶ Explain how air pollution can be reduced. SAI 1, SPSP 1–5, ES 1k, 2a	**Demonstration,** p. 23 in ATE **Interactive Explorations CD-ROM,** Moose Malady *A **Worksheet** is also available in the **Interactive Explorations Teacher's Edition.*** **Long-Term Projects & Research Ideas,** A Breath of Fresh Ether?

*See page **T23** for a complete correlation of this book with the*

NATIONAL SCIENCE EDUCATION STANDARDS.

TECHNOLOGY RESOURCES

 Guided Reading Audio CD
English or Spanish, Chapter 1

One-Stop Planner CD-ROM with Test Generator

 **Interactive Explorations CD-ROM**
CD 2, Exploration 3, Moose Malady

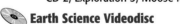 **Earth Science Videodisc**
Global Winds: 29493–33634
Local Winds: 33668–36238

 CNN. **Eye on the Environment,** Global Warming, Segment 13
CO_2 and Arctic Ozone, Segment 14

Multicultural Connections, China Coal, Segment 3

Scientists in Action, Tracking Mercury in the Everglades, Segment 15

CLASSROOM WORKSHEETS, TRANSPARENCIES, AND RESOURCES	SCIENCE INTEGRATION AND CONNECTIONS	REVIEW AND ASSESSMENT
Directed Reading Worksheet **Science Puzzlers, Twisters & Teasers**		
Directed Reading Worksheet, Section 1 **Transparency 13,** Photosynthesis and Respiration: What's the Connection? **Transparency 162,** Profile of the Earth's Atmosphere **Reinforcement Worksheet,** Earth's Amazing Atmosphere	**Chemistry Connection,** p. 5 **Connect to Life Science,** p. 5 in ATE **Apply,** p. 7 **Connect to Physical Science,** p. 8 in ATE **Multicultural Connection,** p. 8 in ATE	**Self-Check,** p. 6 **Section Review,** p. 9 **Quiz,** p. 9 in ATE **Alternative Assessment,** p. 9 in ATE
Directed Reading Worksheet, Section 2 **Transparency 163,** Radiation and the Atmosphere **Transparency 164,** Radiation, Convection, and Conduction **Transparency 165,** The Greenhouse Effect	**Connect to Environmental Science,** p. 10 in ATE **Biology Connection,** p. 13 **Connect to Life Science,** p. 13 in ATE	**Homework,** p. 11 in ATE **Section Review,** p. 13 **Quiz,** p. 13 in ATE **Alternative Assessment,** p. 13 in ATE
Directed Reading Worksheet, Section 3 **Transparency 166,** Sea and Land Breezes	**Connect to Physical Science,** p. 14 in ATE **Multicultural Connection,** p. 15 in ATE **Environment Connection,** p. 17 **Connect to Life Science,** p. 17 in ATE **Math and More,** p. 18 in ATE **Multicultural Connection,** p. 18 in ATE **MathBreak,** Calculating Groundspeed, p. 19	**Homework,** p. 17 in ATE **Section Review,** p. 19 **Quiz,** p. 19 in ATE **Alternative Assessment,** p. 19 in ATE
Directed Reading Worksheet, Section 4 **Transparency 167,** The Formation of Smog **Critical Thinking Worksheet,** The Extraordinary GBG5K	**Connect to Physical Science,** p. 22 in ATE **Multicultural Connection,** p. 22 in ATE **Real-World Connection,** p. 22 in ATE **Cross-Disciplinary Focus,** p. 24 in ATE **Health Watch:** Particles in the Air, p. 32 **Scientific Debate:** A Cure for Air Pollution? p. 33	**Section Review,** p. 25 **Quiz,** p. 25 in ATE **Alternative Assessment,** p. 25 in ATE

 internet **connect**

 go.hrw .com **Holt, Rinehart and Winston On-line Resources**

go.hrw.com

For worksheets and other teaching aids related to this chapter, visit the HRW Web site and type in the keyword: **HSTATM**

 SCI LINKS NSTA **National Science Teachers Association**

www.scilinks.org

Encourage students to use the *sci*LINKS numbers listed in the internet connect boxes to access information and resources on the **NSTA** Web site.

END-OF-CHAPTER REVIEW AND ASSESSMENT

Chapter Review in Study Guide

Vocabulary and Notes in Study Guide

Chapter Tests with Performance-Based Assessment, Chapter 1 Test

Chapter Tests with Performance-Based Assessment, Performance-Based Assessment 1

Concept Mapping Transparency 15

Chapter Resources & Worksheets

Visual Resources

TEACHING TRANSPARENCIES

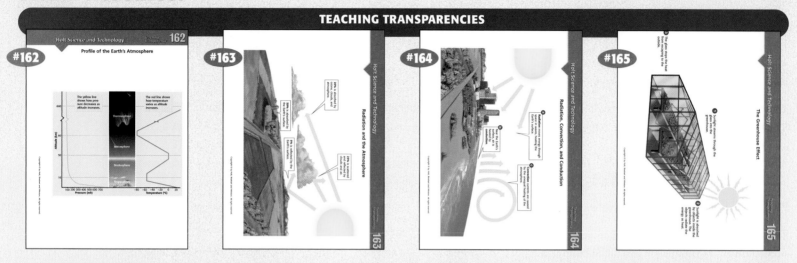

#162 — Holt Science and Technology — Profile of the Earth's Atmosphere — 162

#163 — Holt Science and Technology — Radiation and the Atmosphere — 163

#164 — Holt Science and Technology — Radiation, Convection, and Conduction — 164

#165 — Holt Science and Technology — The Greenhouse Effect — 165

TEACHING TRANSPARENCIES

CONCEPT MAPPING TRANSPARENCY

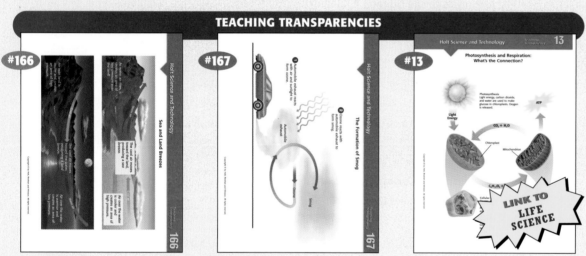
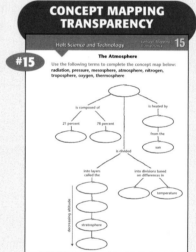

#166 — Holt Science and Technology — Sea and Land Breezes — 166

#167 — Holt Science and Technology — The Formation of Smog — 167

#13 — Holt Science and Technology — Photosynthesis and Respiration: What's the Connection? — 13

#15 — Holt Science and Technology — The Atmosphere — Use the following terms to complete the concept map below: radiation, pressure, mesosphere, atmosphere, nitrogen, troposphere, oxygen, thermosphere — 15

Meeting Individual Needs

DIRECTED READING

#1 — DIRECTED READING WORKSHEET — *The Atmosphere*

Chapter Introduction
As you begin this chapter, answer the following.
1. Read the title of the chapter. List three things that you already know about this subject.

2. Write two questions about this subject that you would like answered by the time you finish this chapter.

3. How does the title of the Start-Up Activity relate to the subject of the chapter?

Section 1: Characteristics of the Atmosphere (p.4)
4. Why is the atmosphere important to us? (Circle all that apply.)
   a. It contains the oxygen we breathe.
   b. It keeps the clouds close to the Earth.
   c. It protects us from the sun's harmful rays.
   d. It holds onto the Earth's surface.

REINFORCEMENT & VOCABULARY REVIEW

#1 — REINFORCEMENT WORKSHEET — *Earth's Amazing Atmosphere*

Complete this worksheet after you finish reading Chapter 15, Section 1.
The Earth's atmosphere is divided into four layers. Use the directions below to label the diagram of the Earth's atmosphere on the next page. Match the layer of the Earth's atmosphere with its description, and write the corresponding letter in the space provided.

Column A	Column B
___ 1. The layer of the Earth's atmosphere that you live in.	a. troposphere
___ 2. The coldest layer of the Earth's atmosphere. This layer is directly below the uppermost layer.	b. stratosphere
___ 3. The uppermost layer of the atmosphere.	c. mesosphere
___ 4. The layer that contains most of the atmosphere's ozone. It is the layer above the layer that you live in.	d. thermosphere

5. Use the descriptions above to label the four layers on the diagram on the next page.
6. There is no clear boundary between the uppermost layer of the atmosphere and space. The atmosphere becomes thinner and thinner and blends into space. At the very top of the diagram, write in the word space with an arrow pointing up.
7. The ozone layer is in the upper part of the atmospheric layer that contains most of the atmosphere's ozone. Draw in the ozone layer on the diagram.
8. The ozone layer is important because it absorbs ultraviolet radiation. Draw a wavy line coming from space to represent the UV radiation that is absorbed by the ozone layer.
9. The thermosphere contains ions, which are electrically charged particles. Ions of nitrogen and oxygen atoms absorb solar energy, they become ions. Draw in ions in the thermosphere. Remember that the thermosphere is very thin, and there will be almost no ions near the top of this atmospheric layer.
10. The troposphere is the densest layer of the atmosphere. It is much denser than the other layers. Shade this layer heavily to indicate how dense its air is.
11. The stratosphere is very thin. Shade this layer lightly.
12. The mesosphere is even less dense than the stratosphere. Shade this layer very lightly.

#1 — VOCABULARY REVIEW WORKSHEET — *In the Air*

After you finish reading Chapter 15, try this crossword puzzle!
Use the clues below to complete the crossword puzzle on the next page.

ACROSS
2. atmospheric layer above the troposphere
4. height of an object above the Earth's surface
7. the coldest layer of the atmosphere
8. Pollutants such as ozone or smog are _____ pollutants.
11. the effect that causes objects to move in a curved direction due to the Earth's rotation
12. a device used to remove pollutants from smokestacks
15. wind belts that extend from the poles to 60° latitude
16. molecule made up of three oxygen atoms
20. moving air
21. Pollutants in the air because of human activity are _____ pollutants.
22. the effect in which gases in the atmosphere convert absorbed radiation into heat
23. heat transfer from one material to another by direct contact

DOWN
1. narrow belts of high-speed winds
3. winds that blow from 30° latitude to the equator
6. the uppermost atmospheric layer
9. mixture of gases that surrounds the Earth
9. the measure of the force with which the air molecules are pushing on a surface
13. a rise in average global temperatures
14. movement of heat by a liquid or gas
14. global winds found between 30° and 60° latitude
16. damaging type of precipitation caused by oxides of sulfur and nitrogen
17. the layer of the atmosphere where we live
19. energy that travels in waves

SCIENCE PUZZLERS, TWISTERS & TEASERS

#1 — SCIENCE PUZZLERS, TWISTERS & TEASERS — *The Atmosphere*

Some Like It Hot
1. Andrea likes school, but she loves the summer even more. Over the summer, she observed many processes that reminded her of energy concepts she learned about in school. Fill in the energy concepts she has observed in the space provided. Choose from radiation, conduction, convection, greenhouse effect, and global warming.
   a. Doesn't it seem like lately every summer is the hottest on record?

   b. Ouch! The send is burning my feet! Why didn't I bring my sandals?

   c. Ugh! Roll down the window! Next time we should park in the shade. I'm melting back here.

   d. This is my favorite part of making soup. See how the spices come up in the middle of the pot then go shooting out to the side and back down? Over and over. That's so cool.

   e. I love to be outside. The sun makes me feel warm all over.

Blow Wind Blow
2. Identify the different types of winds from the clues given below.
   a. When these winds failed, early traders would give just about anything to get them back.

   b. They will take you back home if you're European.

   c. They ruffle the penguin's feathers and the polar bear's fur.

   d. Airline pilots use these high winds to go with the flow whenever they can.

Review & Assessment

STUDY GUIDE

#1 VOCABULARY & NOTES WORKSHEET
The Atmosphere

By studying the Vocabulary and Notes listed for each section below, you can gain a better understanding of this chapter.

SECTION 1
Vocabulary
In your own words, write a definition of each of the following terms in the space provided.

1. atmosphere

2. air pressure

3. altitude

4. troposphere

5. stratosphere

6. ozone

7. mesosphere

8. thermosphere

Notes
Read the following section highlights. Then, in your own words, write the highlights in your ScienceLog.
• The atmosphere is a mixture of gases.
• Nitrogen and oxygen are the two most abundant atmospheric gases.

#1 CHAPTER REVIEW WORKSHEET
The Atmosphere

USING VOCABULARY
For each pair of terms, explain the difference in their meanings.

1. air pressure/altitude

2. troposphere/thermosphere

3. greenhouse effect/global warming

4. convection/conduction

5. global wind/local wind

CHAPTER TESTS WITH PERFORMANCE-BASED ASSESSMENT

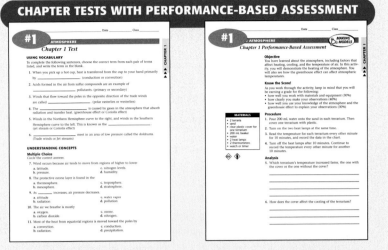

#1 ATMOSPHERE
Chapter 1 Test

USING VOCABULARY
To complete the following sentences, choose the correct term from each pair of terms listed, and write the term in the blank.

1. When you pick up a hot cup, heat is transferred from the cup to your hand primarily by _____. (conduction or convection)

2. Acids formed in the air from sulfur compounds are an example of _____ pollutants. (primary or secondary)

3. Winds that flow toward the poles in the opposite direction of the trade winds are called _____. (polar easterlies or westerlies)

4. The _____ is caused by gases in the atmosphere that absorb radiation and transfer heat. (greenhouse effect or Coriolis effect)

5. Winds in the Northern Hemisphere curve to the right, and winds in the Southern Hemisphere curve to the left. This is known as the _____. (jet stream or Coriolis effect)

6. _____ meet in an area of low pressure called the doldrums. (Trade winds or Jet streams)

UNDERSTANDING CONCEPTS
Multiple Choice
Circle the correct answer.

7. Wind occurs because air tends to move from regions of higher to lower
   a. latitude.
   b. pressure.
   c. nitrogen levels.
   d. humidity.

8. The protective ozone layer is found in the
   a. thermosphere.
   b. mesosphere.
   c. troposphere.
   d. stratosphere.

9. As _____ increases, air pressure decreases.
   a. altitude
   b. radiation
   c. water vapor
   d. pollution

10. The air we breathe is mostly
    a. oxygen.
    b. carbon dioxide.
    c. ozone.
    d. nitrogen.

11. Most of the heat from equatorial regions is moved toward the poles by
    a. convection.
    b. radiation.
    c. conduction.
    d. precipitation.

#1 THE ATMOSPHERE — MAKING MODELS
Chapter 1 Performance-Based Assessment

Objective
You have learned about the atmosphere, including factors that affect heating, cooling, and the temperature of air. In this activity, you will demonstrate the heating of the atmosphere. You will also see how the greenhouse effect can affect atmospheric temperatures.

Know the Score!
As you work through the activity, keep in mind that you will be earning a grade for the following:
• how well you work with materials and equipment (30%)
• how clearly you make your observations (40%)
• how well you use your knowledge of the atmosphere and the greenhouse effect to explain your observations (30%)

MATERIALS
• 2 terraria
• sand
• clear plastic cover for one terrarium
• 200 mL beaker
• water
• 2 heat lamps
• 2 thermometers
• watch or timer

Procedure
1. Pour 200 mL water onto the sand in each terrarium. Then cover one terrarium with plastic.
2. Turn on the two heat lamps at the same time.
3. Read the temperature for each terrarium every other minute for 10 minutes, and record the data in the chart.
4. Turn off the heat lamps after 10 minutes. Continue to record the temperature every other minute for another 10 minutes.

Analysis
5. Which terrarium's temperature increased faster, the one with the cover or the one without the cover?

6. How does the cover affect the cooling of the terrarium?

Lab Worksheets

ECOLABS & FIELD ACTIVITIES

#1 STUDENT WORKSHEET — MAKING MODELS
That Greenhouse Effect!

Welcome to another round of *That Greenhouse Effect!*—the game show on which the contestants not only predict outcomes but also use their keen intellect while working against the clock. I am your host Blaise Haht. Today, contestants are warming up to investigate the results of the greenhouse effect. First let's introduce the contestants.

Meet Professor Luke Wharm, who will determine whether the air above land surfaces or the air above dirt surfaces is cooler. At the water versus land station is Ms. Sylvia Aguapher, two-time medalist in swimming. Mr. Phil Buentot, a landscape architect, is at the wet-soil-versus-dry-soil station. Mr. Ed Blooms, a local paleobotanist, will be at the plants-versus-no-plants station. And last but not least, Ms. Lilith Friese, a professional ice-wall climber, will test the effects of ice and snow on the greenhouse effect.

Each contestant will have 40 minutes to construct two greenhouse models and determine which condition is cooler. Let's get the game going. Grab a jar, and check out the heat!

MATERIALS
• 2 index cards
• transparent tape
• 2 thermometers
• 2 large glass jars of the same size and shape
• clear plastic wrap
• 2 rubber bands
• watch or clock
• one or more of the following items: black and white latex paint, water, soil, sod, plants, ice cubes
• straightedge or ruler
• graph paper

Ask a Question
How do different surface conditions contribute to the temperature of Earth's atmosphere?

Make a Prediction: Round 1
1. In the first round, you have 20 minutes to construct a model and determine what would happen if sunlight were trapped in Earth's atmosphere. Write your prediction in your ScienceLog.

Conduct an Experiment
2. **Inside:** Tape an index card around the bulb of each thermometer to shield the bulb from the sun.
3. Tape each thermometer inside a jar so that the bulb doesn't touch the bottom of the jar, as shown at right.
4. In your ScienceLog, draw a table like the one on page 58. In your data table, record the initial temperature of both thermometers.
5. Cover the opening of one jar with plastic wrap. Secure the wrap with a rubber band. This will model the greenhouse effect on Earth while the uncovered jar serves as a control.
6. **Outside:** Place the jars in a bright, sunny spot. Read both thermometers every 2 minutes for 10 minutes. Record the temperatures in your data table.

WHIZ-BANG DEMONSTRATIONS

#1 TEACHER-LED DEMONSTRATION — DISCOVERY LAB
Blue Sky

Purpose
Students investigate why the midday sky is blue and why sunsets are red.

Time Required
10–15 minutes

Lab Ratings
Teacher Prep
Concept Level
Clean Up

MATERIALS
• empty aquarium
• tap water
• flashlight or slide projector
• powdered milk or powdered coffee creamer
• blank white card

Advance Preparation
Prepare a solution as follows to simulate the atmosphere.
• Fill the aquarium with water, and shine a beam of light into the water so that the beam is parallel to the length of the aquarium.
• Stir a small amount of powdered milk or powdered coffee creamer into the water. As you do so, the beam of light will become more noticeable.
• Continue to add powder until the beam is clearly visible from across the room. Do not add too much powder! Remove the beam of light.

Before performing this demonstration for the class, you may wish to turn out the lights and allow time for everyone's eyes to adjust to the darkness.

HELPFUL HINT
If students have difficulty seeing the colors, try dimming the lights in the classroom further or using a narrower beam of light. One way to produce a narrow beam is to use a hole punch to create a hole in an unexposed black slide or in an index card cut to the size of a slide. Then place the slide or index card in a slide projector and focus the projector to obtain a sharp, narrow beam.

What to Do
1. Ask students to offer an explanation of why the sky is often blue at midday and orange-red at sunset. (Accept all reasonable responses.) Tell them that you will now perform a demonstration to help them determine the answers.
2. Shine the beam of light through the tank from one end of the aquarium. Encourage a few students at a time to walk around the tank and observe the beam from different positions.
3. Ask a volunteer to hold up a white card at the other end of the beam in the path of the beam. This will allow students to observe the color of the light as it leaves the water.
4. After students have made their initial observations, add more powder to the water. Students should observe that the colors along the beam change from blue-white to yellow-orange.
5. Offer students a chance to revise their previous explanations based on what they have just observed.

continued...

LONG-TERM PROJECTS & RESEARCH IDEAS

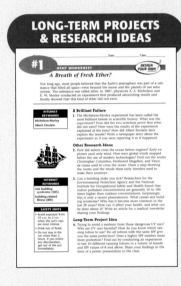

#1 STUDENT WORKSHEET — DESIGN YOUR OWN
A Breath of Fresh Ether?

Not long ago, most people believed that the Earth's atmosphere was part of a substance that filled all space—even beyond the moon and the planets of our solar system. The substance was called ether. In 1887, physicists A. A. Michelson and E. W. Morley conducted an experiment that produced astonishing results and finally showed that this kind of ether did not exist.

INTERNET KEYWORDS
Michelson-Morley
Albert Einstein

A Brilliant Failure
1. The Michelson-Morley experiment has been called the most brilliant failure in scientific history. What was the experiment? How did the two scientists prove that ether did not exist? How were the results of the experiment explained at the time? How did Albert Einstein later explain the results? Write a newspaper story about the experiment as if you were reporting it as it happened.

Other Research Ideas
2. How did sailors cross the ocean before engines? Early explorers used only wind. How were global winds mapped before the use of modern technologies? Find out the routes Christopher Columbus, Ferdinand Magellan, and Vasco da Gama used to cross the ocean. Draw a map showing the routes and the winds these early travelers used to make their journeys.

INTERNET KEYWORDS
sick building syndrome (SBS)
building-related illness (BRI)

SAFETY HINTS
• Avoid exposure from 10 A.M. to 2 P.M. when the sun's rays are most intense.
• Drink lots of fluids.
• Do not stay in the sun more than 2 hours. If you notice any discoloration, get out of the sun immediately.

3. Can a building make you sick? Researchers for the Environmental Protection Agency and the National Institute for Occupational Safety and Health found that indoor pollutant concentrations are generally 10 to 100 times higher than outdoor concentrations. Surprisingly, this is only a recent phenomenon. What causes sick building syndrome? Why has it become more common in the last 20 years? How can it affect your health, and what can be done about it? Write an article for a medical newsletter sharing your findings.

Long-Term Project Idea
4. Trying to avoid a sunburn from those dangerous UV rays? Why are UV rays harmful? How do you know which tanning lotion to use? Do all lotions with the same SPF give you the same protection? Does a higher SPF number mean more protection? Find out by conducting an experiment to test 10 different tanning lotions in a variety of brands and SPF values of 8 and above. Share your findings in the form of a poster presentation to the class.

DATASHEETS FOR LABBOOK

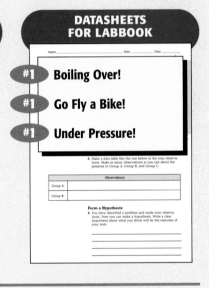

#1 Boiling Over!
#1 Go Fly a Bike!
#1 Under Pressure!

3. Make a data table like the one below to list your observations. Make as many observations as you can about the potatoes in Group A, Group B, and Group C.

Observations
Group A:
Group B:

Form a Hypothesis
4. You have identified a problem and made your observations. Now you can make a hypothesis. Write a clear hypothesis about what you think will be the outcome of your tests.

Applications & Extensions

CRITICAL THINKING & PROBLEM SOLVING

#1 CRITICAL THINKING WORKSHEET
The Extraordinary GBG5K

While standing in line at the local Sack-N-Spend, you spot a tabloid newspaper. Next to the headline "Man Gives Birth to Baby Seal," you read this headline:

– SCIENTIST FINDS AIR POLLUTION SOLUTION –

Because you've been studying the atmosphere in school, you buy the paper to impress your teacher with your new knowledge. Here's a clip from the article:

Bolognaville, Antarctica: Charlie Bloomfritter, "Earth scientist extraordinaire," has invented a surefire way to end air pollution.

Says Mr. Bloomfritter, "With my patented Grunk-B-Gone 5000 filtration machine, we can stop worrying about air pollution. As you can see, all solid particles in the air that go into the GBG5K do not come out! They are trapped inside forever. I have plans to build a super-large version to place in the parking lot outside, which will be so powerful that it will filter all the gunk out of the Earth's atmosphere. People can now pollute all they want because the GBG5K will clean up the mess."

Bloomfritter is a self-educated "gunkologist." He wrote the article, "Building a Better Time Machine," in last week's issue.

Evaluating Sources
1. Does this publication seem like a reliable source of scientific facts? Explain.

Reading for Detail
2. What kinds of pollutants does Bloomfritter claim the GBG5K will catch? Give at least two examples.

MULTICULTURAL CONNECTIONS

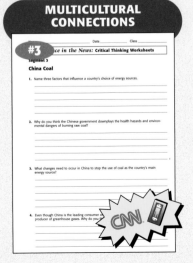

#3 Science in the News: Critical Thinking Worksheets
Segment 3
China Coal

1. Name three factors that influence a country's choice of energy sources.

2. Why do you think the Chinese government downplays the health hazards and environmental dangers of burning raw coal?

3. What changes need to occur in China to stop the use of coal as the country's main energy source?

4. Even though China is the leading consumer of coal, it is not the leading producer of greenhouse gases. Why do you... [CNN]

EYE ON THE ENVIRONMENT

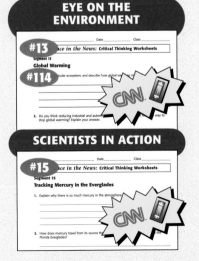

#13 Science in the News: Critical Thinking Worksheets
Segment 13
Global Warming

#114

... particular ecosystem, and describe how global warming... way to... [CNN]

2. Do you think reducing industrial and automotive... stop global warming? Explain your answer.

SCIENTISTS IN ACTION

#15 Science in the News: Critical Thinking Worksheets
Segment 15
Tracking Mercury in the Everglades

1. Explain why there is so much mercury in the atmosphere.

2. How does mercury travel from its source to the Florida Everglades? [CNN]

INTERACTIVE EXPLORATIONS

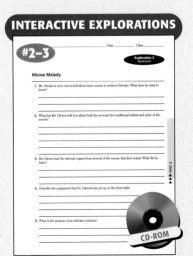

#2-3 — Exploration 3 Worksheet
Moose Malady

1. Mr. Oleson is very concerned about some moose in western Sweden. What does he want to know?

2. What has Mr. Oleson told you about both the new and the traditional habitat and niche of the moose?

3. Mr. Oleson had the internal organs from several of the moose that died tested. What did he learn?

4. Describe the equipment that Dr. Labcoat has set up on the front table.

5. What is the purpose of an indicator solution?

CD-ROM

SECTION 1

Characteristics of the Atmosphere

▶ **Take a Deep Breath!**
Near the Earth's surface, the atmosphere consists of 78.08 percent nitrogen, 20.95 percent oxygen, 0.93 percent argon, 0.03 percent carbon dioxide, and traces of water vapor. Earth's early atmosphere was quite different than it is today, consisting of about 79 percent water vapor, 11–12 percent carbon dioxide, 6 percent sulfur dioxide, 1 percent nitrogen, less than 1 percent hydrogen, and traces of other gases. Scientists theorize that about 95 percent of the oxygen present in today's atmosphere formed as a byproduct of photosynthesis.

▶ **Radio Days**
Fadeouts of radio communications are due to sudden ionospheric disturbances, known as SIDs. These storms can last 15 to 30 minutes and are caused by violent solar outbursts that release electrically charged particles.

IS THAT A FACT!

➤ The Earth's troposphere contains almost 90 percent of the atmosphere's total mass. In the troposphere, temperature decreases at an average rate of 6.4°C/km with increasing altitude.

SECTION 2

Heating of the Atmosphere

▶ **Cloudy with a Chance of . . .**
Clouds reflect incoming solar radiation very effectively. The amount of solar radiation a cloud can reflect depends on its thickness. A cloud less than 50 m thick can reflect up to 40 percent of incoming solar radiation, while a cloud more than 5,000 m thick can reflect 80 percent or more. The average reflectivity of clouds is about 55 percent.

▶ **Greenhouse Gases**
Water vapor, carbon dioxide, ozone, methane, and chlorofluorocarbons are often called greenhouse gases. These gases transmit incoming short-wave radiation from the sun and absorb much of the outgoing long-wave radiation from the Earth's surface.

▶ **Global Warming—An Idea Before Its Time!**
Since the 1970s, global warming has been a topic of concern. However, a global warming model was proposed as early as 1896 by a Swedish physicist and chemist named Svante Arrhenius. Arrhenius theorized that the carbon dioxide released from burning coal would increase the intensity of Earth's greenhouse effect and lead to global warming. In 1954, it was first suggested that deforestation increases the amount of CO_2 in the atmosphere. Since then, numerous scientific studies have examined the effects of carbon dioxide on the temperature of Earth's atmosphere.

SECTION 3

Atmospheric Pressure and Winds

▶ **Gustave Coriolis**
Gustave Gaspard Coriolis was a French mathematician and engineer who lived and worked in Paris from

1792 to 1843. His most well-known contribution to science is a paper published in 1835 that introduces the Coriolis force. In "On the Equations of Relative Motion of Systems of Bodies," Coriolis proves that an inertial force (the Coriolis force) acts on a rotating object at a right angle to the object's motion. The Coriolis force causes matter to be deflected from its original path. This force influences the general direction of global winds and open-ocean circulation, as well as the rotational movements of severe weather, such as hurricanes.

IS THAT A FACT!

- ☛ When airplanes fly north or south, pilots have to make corrections to counteract the Coriolis effect.

▶ Jet Streaks

Jet streaks are winds within jet streams that flow faster than the adjacent winds. Jet streaks influence storm formation and associated precipitation. Rising jet streaks and the low-pressure area that forms beneath them present favorable conditions for storms to form. Sinking jet streaks inhibit storm formation and precipitation.

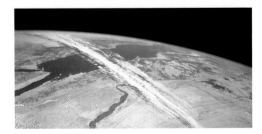

IS THAT A FACT!

- ☛ The speed of jet streams ranges from about 92 km/h to more than 483 km/h!

The Air We Breathe

▶ Vog and Laze

Two pollution problems associated with volcanic activity at Earth's surface are vog and laze. Vog is volcanic smog that forms when the sulfur dioxide released during an eruption reacts with sunlight, dust particles, water vapor, and oxygen to form various sulfur compounds, including sulfur dioxide. When Kilauea began to erupt in 1983, it sent about 2,000 tons of sulfur dioxide into the air each day.

- Laze is lava haze, a form of pollution that forms when lava flows react with ocean water. The extreme temperature of the lava causes sea water to vaporize. Physical and chemical interactions create white plumes of hydrochloric acid and concentrated salt water that can threaten the health of people living near volcanoes.

▶ Allowance Trading System

An important part of the EPA's Acid Rain Program is the allowance trading system, which is designed to reduce sulfur-dioxide emissions. In this system, 1 ton of sulfur dioxide (SO_2) emission is equivalent to one allowance. There are a limited number of allowances allocated for each year. Companies can buy, sell or trade allowances freely, but if they exceed their allowances, they must pay a punitive fine. The system allows a company to determine the most cost-effective ways to comply with the Clean Air Act. It can reduce emissions by using technology that conserves energy, by using renewable energy sources, or by updating its pollution-control devices and using low-sulfur fuels.

- Sulfur dioxide allowances are inexpensive and can be bought from the EPA by private citizens, schools, and community and environmental organizations. Purchasing one allowance will reduce a company's allowable SO_2 emissions in a given year by 1 ton.

For background information about teaching strategies and issues, refer to the *Professional Reference for Teachers*.

CHAPTER

1

The Atmosphere

Sections

 Pre-Reading
Questions

1. What is air made of?
2. How is the atmosphere organized?
3. What is wind and how does it move?

2

FLOATING ON AIR

These skydivers might have checked their parachutes at least a half dozen times before they jumped. They probably also paid particular attention to the day's weather report. Skydivers should know what to expect from the atmosphere. The atmosphere can be unpredictable and dangerous, but it also provides us with the gases needed for our survival on Earth. In this chapter, you will learn about the Earth's atmosphere and how it affects your life.

---

internetconnect

HRW On-line Resources	SCiLINKS NSTA	Smithsonian Institution	CNNfyi.com
go.hrw.com	**www.scilinks.com**	**www.si.edu/hrw**	**www.cnnfyi.com**
For worksheets and other teaching aids, visit the HRW Web site and type in the keyword: **HSTATM**	Use the *sci*LINKS numbers at the end of each chapter for additional resources on the **NSTA** Web site.	Visit the Smithsonian Institution Web site for related on-line resources.	Visit the CNN Web site for current events coverage and classroom resources.

START-UP Activity

AIR—IT'S MASSIVE

In this activity, you will find out if air has mass.

Procedure

1. Use a **scale** to find the mass of a **ball,** such as a football or a basketball, with no air in it. Record the mass of the empty ball in your ScienceLog.

2. Pump up the ball with an **air pump.**

3. Use the scale to find the mass of the ball filled with air. Record the mass of the ball filled with air in your ScienceLog.

Analysis

4. Compare the mass of the empty ball with the mass of the ball filled with air. Did the mass of the ball change after you pumped it up?

5. Based on your results, does air have mass? Explain your answer.

3

START-UP Activity

AIR—IT'S MASSIVE

MATERIALS
FOR EACH GROUP:
• scale
• football or basketball, deflated
• air pump

Answers to START-UP Activity

4. The mass of the empty ball is less than the mass of the ball filled with air. The mass of the ball increased when the ball was pumped up.

5. Air has mass as shown by the increase in mass when the ball is filled with air.

Focus

Characteristics of the Atmosphere

This section defines the atmosphere and explains its basic characteristics. It discusses the atmosphere's composition and explains how temperature and pressure are related to altitude. The section concludes with a description of the four layers of the Earth's atmosphere.

🔔 Bellringer

Have students list the ways that the atmosphere is different from outer space in their ScienceLog. Tell students that a little more than a century ago, many scientists believed that the Earth's atmosphere blended with a hypothetical substance called *ether* that filled the entire universe. In 1887, the physicist A. A. Michelson demonstrated that the universe is not filled with ether.

1️⃣ Motivate

DEMONSTRATION

Air Pressure Fill a paper cup with water, and push a square piece of cardboard firmly against the cup's mouth with one hand. Hold the cardboard in place as you position the cup over a volunteer's hand. Ask the class to predict what will happen when you invert the cup. Quickly invert the cup, being careful to keep the cardboard in place. The cardboard should stay in place, and the water should stay in the cup. Explain that this demonstration illustrates that the atmosphere exerts pressure in all directions.

Note: Practice this demonstration over a sink before doing it in front of the class.

Terms to Learn

atmosphere	stratosphere
air pressure	ozone
altitude	mesosphere
troposphere	thermosphere

What You'll Do

◆ Discuss the composition of the Earth's atmosphere.
◆ Explain why pressure changes with altitude.
◆ Explain how temperature changes with altitude.
◆ Describe the layers of the atmosphere.

Characteristics of the Atmosphere

If you were lost in the desert, you could survive for a few days without food and water. But you wouldn't last more than 5 minutes without the *atmosphere*. The **atmosphere** is a mixture of gases that surrounds the Earth. In addition to containing the oxygen we need to breathe, it protects us from the sun's harmful rays. But the atmosphere is always changing. Every breath we take, every tree we plant, and every motor vehicle we ride in affects the composition of our atmosphere. Later you will find out how the atmosphere is changing. But first you need to learn about the atmosphere's composition and structure.

Composition of the Atmosphere

Figure 1 shows the relative amounts of the gases that make up the atmosphere. Besides gases, the atmosphere also contains small amounts of solids and liquids. Tiny solid particles, such as dust, volcanic ash, sea salt, dirt, and smoke, are carried in the air. Next time you turn off the lights at night, shine a flashlight and you will see some of these tiny particles floating in the air. The most common liquid in the atmosphere is water. Liquid water is found as water droplets in clouds. Water vapor, which is also found in the atmosphere, is a gas and is not visible.

Figure 1 *Two gases—nitrogen and oxygen—make up 99 percent of the air we breathe.*

Nitrogen is the most abundant gas in the atmosphere. It is released into the atmosphere by volcanic eruptions and when dead plants and dead animals decay.

Oxygen, the second most common gas in the atmosphere, is produced mainly by plants.

The **remaining 1 percent** of the atmosphere is made up of argon, carbon dioxide, water vapor, and other gases.

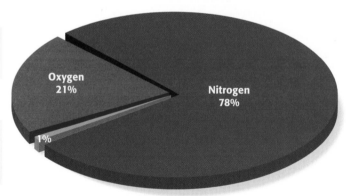

Oxygen 21%

Nitrogen 78%

1%

internetconnect

SCLINKS **NSTA**
TOPIC: Composition of the Atmosphere
GO TO: www.scilinks.org
*sci*LINKS **NUMBER:** HSTE355

Directed Reading Worksheet Section 1

🧪 WEIRD SCIENCE

An experiment in 1664 demonstrated the force exerted by the Earth's atmosphere. Most of the air was removed from a hollow sphere whose halves had been sealed together with an airtight gasket. Sixteen horses were needed to pull the metal hemispheres apart!

Atmospheric Pressure and Temperature

Have you ever been in an elevator in a tall building? If you have, you probably remember the "popping" in your ears as you went up or down. As you move up or down in an elevator, the air pressure outside your ears changes, while the air pressure inside your ears stays the same. **Air pressure** is the measure of the force with which the air molecules push on a surface. Your ears pop when the pressure inside and outside of your ears suddenly becomes equal. Air pressure changes throughout the atmosphere. Temperature and the kinds of gases present also change. Why do these changes occur? Read on to find out.

Pressure Think of air pressure as a human pyramid, as shown in **Figure 2.** The people at the bottom of the pyramid can feel all the weight and pressure of the people on top. The person on top doesn't feel any weight because there isn't anyone above. The atmosphere works in a similar way.

The Earth's atmosphere is held around the planet by gravity. Gravity pulls the gas molecules in the atmosphere toward the Earth's surface, giving them weight. This weight causes the air to push against the Earth's surface. As you move farther away from the Earth's surface, air pressure decreases because fewer gas molecules are pushing on you. **Altitude** is the height of an object above the Earth's surface. As altitude increases, air pressure decreases.

Chemistry
CONNECTION

Water is the only substance that exists as a liquid, a solid, and a gas in the Earth's atmosphere.

Figure 2 *Like the bottom row of the human pyramid, the lower atmosphere experiences greater pressure than the upper atmosphere.*

READING 📖 STRATEGY

Prediction Guide After students read this page, ask them the following questions:

• Which gas—oxygen or nitrogen—is the major component of Earth's air? (nitrogen)

• Does air contain anything other than gases? (Yes; it contains solids, such as dust, and liquids, such as water.)

GROUP ACTIVITY

It's a Gas! Have small groups demonstrate how oxygen enters the atmosphere. Suggested materials include a freshwater plant, such as *Elodea,* a small plastic storage beaker, a funnel, a test tube, and water. Tell students to immerse the plant in the water-filled beaker and then cover the plant with the inverted funnel. Have them place a water-filled test tube over the funnel's spout and let the setup sit in a well-lighted area for a few days. After this time, students will observe gas bubbles in the test tube. Inform them that the bubbles they see are made of oxygen gas released during the process of photosynthesis. Sheltered English

MISCONCEPTION
/// ALERT \\\

Make sure students realize that water vapor is an invisible gas. The "steam" they observe coming out of a pot of boiling water is composed of water droplets that form as water vapor cools and condenses on particles in the air. Similarly, clouds appear in the sky when the air cools enough for water vapor to condense and form liquid droplets.

CONNECT TO
LIFE SCIENCE

Use Teaching Transparency 13 to show students how photosynthesis and respiration are linked to gas exchange between organisms and the atmosphere. Photosynthesizing plants use carbon dioxide, water, and light energy to produce oxygen. During respiration, plants and animals

consume oxygen and release carbon dioxide, water, and energy.

Teaching Transparency 13 "Photosynthesis and Respiration: What's the Connection?"
LINK TO LIFE SCIENCE

MEETING INDIVIDUAL NEEDS

Learners Having Difficulty
Students might benefit from using a dictionary to learn the prefixes of the following words:

troposphere, stratosphere, mesosphere, thermosphere

Students will learn that *trop-* means "change" or "turn," *strat-* means "layer," *meso-* means "middle," and *therm-* means "heat." Students should also note that *sphere* means "globe" or "ball." Have students use these meanings to remember the layers of the atmosphere.
Sheltered English

USING THE FIGURE

Have students refer to **Figure 3** to answer these questions:

- Which layer of the atmosphere is closest to Earth? (the troposphere)

- How does temperature change within the stratosphere? (For the first few kilometers, the temperature remains fairly constant. Then the temperature begins rising steeply and levels off again toward the top of the layer.)

- Which atmospheric layer has the greatest range of temperatures? (the thermosphere)

- Approximately how thick is the Earth's atmosphere? (about 600 km)

Students may notice that the iridescent cloud in the thermosphere is the aurora borealis and that the white layer at the top of the stratosphere is the ozone layer.

Teaching Transparency 162
"Profile of the Earth's Atmosphere"

Self-Check

Does air become more or less dense as you climb a mountain? Why? *(See page 136 to check your answer.)*

Air Temperature Air temperature also changes as you increase altitude. As you pass through the atmosphere, air temperature changes between warmer and colder conditions. The temperature differences result mainly from the way solar energy is absorbed as it moves downward through the atmosphere. Some parts of the atmosphere are warmer because they contain gases that absorb solar energy. Other parts do not contain these gases and are therefore cooler.

Layers of the Atmosphere

Based on temperature changes, the Earth's atmosphere is divided into four layers—the troposphere, stratosphere, mesosphere, and thermosphere. **Figure 3** illustrates the four atmospheric layers, showing their altitude and temperature. As you can see, each layer has unique characteristics.

Figure 3 Profile of the Earth's Atmosphere

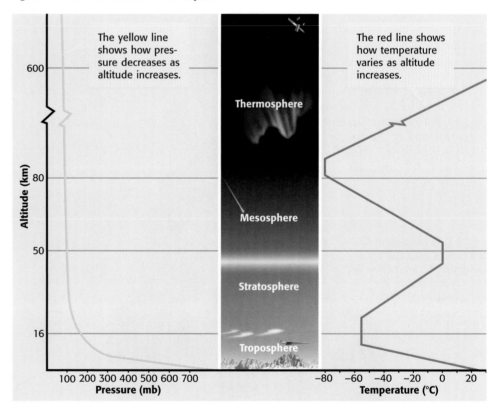

Answer to Self-Check

As you climb a mountain, the air becomes less dense because there are fewer air molecules. So even though cold air is generally more dense than warm air, it is less dense at higher elevations.

IS THAT A FACT!

The oxygen in the Earth's current atmosphere is produced primarily by phytoplankton (tiny, drifting sea plants) and land plants that release oxygen during photosynthesis.

Troposphere The **troposphere**, which lies next to the Earth's surface, is the lowest layer of the atmosphere. The troposphere is also the densest atmospheric layer, containing almost 90 percent of the atmosphere's total mass. Almost all of Earth's carbon dioxide, water vapor, clouds, air pollution, weather, and life-forms are found in the troposphere. In fact, the troposphere is the layer in which you live. **Figure 4** shows the effects of altitude on temperature in the troposphere.

Stratosphere The atmospheric layer above the troposphere is called the **stratosphere.** In the stratosphere, the air is very thin and contains little moisture. The lower stratosphere is extremely cold, measuring about –60°C. In the stratosphere, the temperature rises with increasing altitude. This occurs because of ozone. **Ozone** is a molecule that is made up of three oxygen atoms, as shown in **Figure 5.** Almost all of the ozone in the atmosphere is contained in the *ozone layer* of the stratosphere. Ozone absorbs solar energy in the form of ultraviolet radiation, warming the air. By absorbing the ultraviolet radiation, the ozone layer also protects life at the Earth's surface.

Figure 4 *Snow can remain year-round on a mountain top. That is because as altitude increases, the atmosphere thins, losing its ability to absorb and transfer thermal energy.*

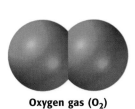 

Oxygen gas (O$_2$)   **Ozone (O$_3$)**

Figure 5 *While ozone is made up of three oxygen atoms, the oxygen in the air you breathe is made up of two oxygen atoms.*

UV and SPFs

People protect themselves from the sun's damaging rays by applying sunblock. Exposure of unprotected skin to the sun's ultraviolet rays over a long period of time can cause skin cancer. The breakdown of the Earth's ozone layer is thinning the layer, which allows some harmful ultraviolet radiation to reach the Earth's surface. Sunblocks contain different ratings of SPFs, or skin protection factors. What do the SPF ratings mean?

MISCONCEPTION ALERT

The amount of stratospheric ozone protecting Earth is smaller than most people realize. If the ozone layer were brought to sea-level pressure and temperature, it would range from 2.5 to 3.5 mm thick! (The ozone layer is thinner above the poles than above the equator.)

Answer to Apply

Answers may vary. Sample answer: SPF indicates how many times longer you can stay in the sun without being burned.

DISCUSSION

Remind students that cold air is more dense than warm air. This is an important fact for pilots to know. Ask students to think about why the Wright brothers tested their biplane early in the morning. (The air is colder and more dense in the morning. Dense air provides more lift to a plane's wings, enabling shorter takeoffs. In addition, controls respond more quickly in the dense air and landing speeds are decreased.)

Ask students to apply this logic to explain why people driving jet-powered rocket cars attempt to break the land-speed record at midday on hot salt flats. (The hot air of the salt flats at midday is not very dense. This reduces drag, allowing the cars to travel nearly 100 mph faster.)

3) Extend

GUIDED PRACTICE

On the board, make a table entitled "The Atmosphere." Include the following headings:

Layer, Pressure range, Temperature range, and Other important information

Have volunteers contribute information for each section of the table. Sheltered English

---

CONNECT TO PHYSICAL SCIENCE

Explain that temperature is a measure of the average kinetic energy of randomly vibrating particles and that heat is the transfer of energy between objects of different temperature. To clarify these concepts, have students imagine a sink full of hot water. Ask them to pretend that they have removed a cup of the hot water from the sink. Students should agree that both volumes of water have the same temperature at this point. Explain that the sink has more thermal energy than the cup because the sink contains more water (and therefore more particles in motion) than the cup.

USING THE FIGURE

Have students use the illustrations in **Figure 6** to compare how energy is transferred in the thermosphere with how it is transferred in the troposphere. How is the thermosphere like a vacuum thermos? (A vacuum thermos surrounds a hot liquid with a partial vacuum. Because there are relatively few air molecules in the vacuum, little energy is transferred from the liquid and it remains hot. Similarly, the thermosphere has few air molecules to transfer energy.)

Large continent-sized wind-storms were detected by the *Upper Atmosphere Research Satellite*. The effect these winds have on weather at the Earth's surface is currently being studied.

Mesosphere Above the stratosphere is the mesosphere. The **mesosphere** is the coldest layer of the atmosphere. As in the troposphere, the temperature drops with increasing altitude. Temperatures can be as low as –93°C at the top of the mesosphere. Scientists have recently discovered large wind storms in the mesosphere with winds reaching speeds of more than 320 km/h.

Thermosphere The uppermost atmospheric layer is the **thermosphere.** Here temperature again increases with altitude because many of the gases are absorbing solar radiation. Temperatures in this layer can reach 1,700°C.

When you think of an area with high temperatures, you probably think of a place that is very hot. While the thermosphere has very high temperatures, it would not feel hot. Temperature and heat are not the same thing. Temperature is a measure of the average energy of particles in motion. A high temperature means that the particles are moving very fast. Heat, on the other hand, is the transfer of energy between objects at different temperatures. But in order to transfer energy, particles must touch one another. **Figure 6** illustrates how the density of particles affects the heating of the atmosphere.

Figure 6 *Temperatures in the thermosphere are higher than those in the troposphere, but the air particles are too far apart for energy to be transferred.*

The **thermosphere** contains few particles that move fast. The temperature of this layer is high due to the speed of its particles. But because the particles rarely touch one another, the thermosphere does not transfer much energy.

The **troposphere** contains more particles that travel at a slower speed. The temperature of this layer is lower than that of the thermosphere. But because the particles are bumping into one another, the troposphere transfers much more energy.

Multicultural CONNECTION

Different cultures have different explanations for the shimmering lights known as the aurora borealis. Inuit groups thought of the aurora borealis as the torches of spirits that guided souls from Earth to paradise. Have students find out about other myths concerning the aurora borealis.

Ionosphere In the upper part of the mesosphere and the lower thermosphere, nitrogen and oxygen atoms absorb harmful solar energy, such as X rays and gamma rays. This absorption not only contributes to the thermosphere's high temperatures but also causes the gas particles to become electrically charged. Electrically charged particles are called ions; therefore, this part of the thermosphere is referred to as the *ionosphere*. Sometimes these ions radiate energy as light of different colors, as shown in **Figure 7.**

Figure 7 *Aurora borealis (northern lights) and aurora australis (southern lights) occur in the ionosphere.*

The ionosphere also reflects certain radio waves, such as AM radio waves. If you have ever listened to an AM radio station, you can be sure that the ionosphere had something to do with how clear it sounded. When conditions are right, an AM radio wave can travel around the world after being reflected off the ionosphere. These radio signals bounce off the ionosphere and are sent back to Earth.

SECTION REVIEW

1. Explain why pressure decreases but temperature varies as altitude increases.

2. What causes air pressure?

3. How can the thermosphere have high temperatures but not feel hot?

4. **Analyzing Relationships** Identify one characteristic of each layer of the atmosphere, and explain how that characteristic affects life on Earth.

internet**connect**

*SCi*LINKS.
NSTA

TOPIC: Composition of the Atmosphere
GO TO: www.scilinks.org
*sci*LINKS NUMBER: HSTE355

4) Close

Quiz

1. What are the two main gases in Earth's atmosphere? (nitrogen and oxygen)

2. What is atmospheric pressure? (Atmospheric pressure is the force exerted by molecules of air on a surface.)

3. Name the layers of the atmosphere, starting with the one closest to Earth. (troposphere, stratosphere, mesosphere, thermosphere)

4. What is the ozone layer, and why is it important to Earth? (The ozone layer is a layer of ozone molecules in the stratosphere. The layer filters ultraviolet radiation from the sun and prevents much of this radiation from reaching Earth.)

5. Explain how density affects energy transfer in the air. (The less dense the air is, the less effective it is at transferring energy. Particles that are farther apart, or less densely packed, are less likely to collide with other particles. Particles must collide with one another in order to transfer energy.)

ALTERNATIVE ASSESSMENT

Writing **Poetry** Have each student write a poem that creatively yet accurately describes one layer of Earth's atmosphere. Allow time for volunteers to read their poem aloud or display the poem for others to read on their own.

PORTFOLIO

Reinforcement Worksheet
"Earth's Amazing Atmosphere"

SECTION 2
READING WARM-UP

Heating of the Atmosphere

Focus

Heating of the Atmosphere

In this section, students learn that the sun is the principal energy source for our planet. They also discover that energy is transferred in one of three ways—by conduction, convection, or radiation. The section concludes with a discussion of the greenhouse effect and global warming.

Bellringer

Have students suppose that they will be vacationing in two unique spots—the Sahara Desert and the Antarctic ice sheet. Have them decide whether white or black clothing would be best for each location. (Because of its high reflectivity, white would be best for the hot desert. Because of its ability to absorb energy, black clothing would be better for the ice sheet.)

1 Motivate

GROUP ACTIVITY

Refer students to **Figure 8,** noting that on average, 5 percent of incoming solar radiation is reflected from Earth's surface and 50 percent is absorbed. Highly reflective surfaces, such as snow, reflect up to 90 percent of solar radiation, while the oceans reflect only 5 percent. This is noticeable in urban areas—cities with few green spaces and a lot of asphalt can have temperatures 10°C higher than surrounding rural areas. This is called the *heat island effect.* Have groups test an experiment to compare the amount of radiation reflected by asphalt, sidewalk, soil, water, and grass.

Terms to Learn

radiation
conduction
convection
greenhouse effect
global warming

What You'll Do

- Describe what happens to radiation that reaches the Earth.
- Summarize the processes of radiation, conduction, and convection.
- Explain how the greenhouse effect could contribute to global warming.

Have you ever walked barefoot across a sidewalk on a sunny day? If so, your foot felt the warmth of the hot pavement. How did the sidewalk become so warm? The sidewalk was heated as it absorbed the sun's energy. The Earth's atmosphere is also heated in several ways by the transfer of energy from the sun. In this section you will find out what happens to the solar energy as it enters the Earth's atmosphere, how the energy is transferred through the atmosphere, and why it seems to be getting hotter every year.

Energy in the Atmosphere

The Earth receives energy from the sun by radiation. **Radiation** is the transfer of energy as electromagnetic waves. Although the sun releases a huge amount of energy, the Earth receives only about two-billionths of this energy. Yet even this small amount of energy has a very large impact on Earth. **Figure 8** shows what happens to all this energy once it enters the atmosphere.

When energy is absorbed by a surface, it heats that surface. For example, when you stand in the sun on a cool day, you can feel the sun's rays warming your body. Your skin

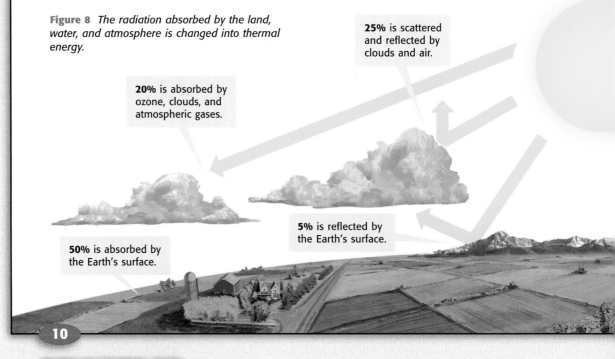

Figure 8 *The radiation absorbed by the land, water, and atmosphere is changed into thermal energy.*

25% is scattered and reflected by clouds and air.

20% is absorbed by ozone, clouds, and atmospheric gases.

5% is reflected by the Earth's surface.

50% is absorbed by the Earth's surface.

10

CONNECT TO ENVIRONMENTAL SCIENCE

The heat island effect occurs because concrete buildings and asphalt absorb solar radiation and reradiate thermal energy, elevating temperatures and increasing the production of smog. The effect is worse in urban areas with little surface water and few trees because the evaporation of water and plant transpiration cools the air.

To counteract the heat island effect, cities have begun to preserve green spaces and plant trees. Some cities are beginning to use construction materials with a higher reflectivity, such as white rooftops and concrete streets. Have groups create a model city in a large box or aquarium. Using two light bulbs and a thermometer, they should test strategies to reduce the heat island effect.

absorbs the radiation, causing your skin's molecules to move faster. You feel this as an increase in temperature. The same thing happens when energy is absorbed by the Earth's surface. The energy from the Earth's surface can then be transferred to the atmosphere, which heats it.

Conduction Conduction is the transfer of thermal energy from one material to another by direct contact. Think back to the example about walking barefoot on a hot sidewalk. Conduction occurs when thermal energy is transferred from the sidewalk to your foot. Thermal energy always moves from warm to cold areas. Just as your foot is heated by the sidewalk, the air is heated by land and ocean surfaces. When air molecules come into direct contact with a warm surface, thermal energy is transferred to the atmosphere.

Convection Most thermal energy in the atmosphere moves by *convection*. **Convection** is the transfer of thermal energy by the circulation or movement of a liquid or gas. For instance, as air is heated, it becomes less dense and rises. Cool air is more dense and sinks. As the cool air sinks, it pushes the warm air up. The cool air is eventually heated by the ground and again begins to rise. This continual process of warm air rising and cool air sinking creates a circular movement of air, called a *convection current,* as shown in **Figure 9.**

BRAIN FOOD

If the Earth is continually absorbing solar energy and changing it to thermal energy, why doesn't the Earth get hotter and hotter? The reason is that much of this energy is lost to space. This is especially true on cloudless nights.

Figure 9 *There are three important processes responsible for heating the Earth and its atmosphere: radiation, conduction, and convection.*

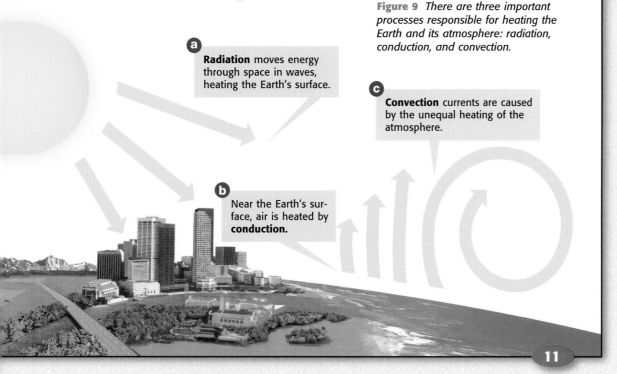

a **Radiation** moves energy through space in waves, heating the Earth's surface.

c **Convection** currents are caused by the unequal heating of the atmosphere.

b Near the Earth's surface, air is heated by **conduction.**

11

Homework

Writing Have students write several paragraphs that compare and contrast various methods used to heat buildings. Suggest that they consider radiators, steam heating systems, heating systems that use furnaces, and solar heating systems. Make sure that students correctly use the terms *radiation, convection,* and *conduction* in their descriptions.

2 Teach

ACTIVITY

Make some popcorn the "old-fashioned" way, using a hot plate or stovetop, oil, and popcorn kernels. As you pop the snack, have volunteers explain how the processes of convection, conduction, and radiation are involved. Point out that a kernel pops when the liquid water stored inside changes to water vapor and expands suddenly. Share the treat with students if time allows. Make sure students with allergies to corn do not eat the popcorn. Sheltered English

DEMONSTRATION

Use heat-resistant gloves, a hot plate, a transparent coffee decanter, water, and confetti to demonstrate convection. Fill the decanter about three-fourths full with water. Sprinkle the confetti into the water, and mix it slightly so that it settles to the bottom. Put on the heat-resistant gloves, and hold the decanter so that only one-half of it is on the hot plate. Heat the water. Have students observe the convection currents move the confetti and explain how this demonstration relates to the movement of air in the atmosphere.

 Directed Reading Worksheet Section 2

 Teaching Transparency 163 "Radiation and the Atmosphere"

 Teaching Transparency 164 "Radiation, Convection, and Conduction"

GROUP ACTIVITY

Model Greenhouses

MATERIALS

FOR EACH GROUP:
• large jar with lid
• thermometer
• small piece of modeling clay

Have students work in small groups to make model greenhouses by placing the thermometer inside the jar and anchoring it with modeling clay. Next have them seal the jar with the lid.

Have each group put its model in a different sunny spot. Students should observe and record changes in temperature every day for 1 week. Students can compare the temperatures they record with the temperatures in a control jar without a lid. Help students infer that solar energy enters a greenhouse and is converted to thermal energy and that the glass prevents most of the thermal energy from escaping.

Sheltered English

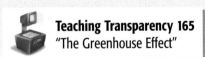 **PG 100**

Boiling Over!

 Teaching Transparency 165
"The Greenhouse Effect"

internet connect

SCiLINKS NSTA

TOPIC: Energy in the Atmosphere
GO TO: www.scilinks.org
*sci*LINKS NUMBER: HSTE360

TOPIC: The Greenhouse Effect
GO TO: www.scilinks.org
*sci*LINKS NUMBER: HSTE365

Annual average surface temperatures in the Northern Hemisphere have been higher in the 1990s than at any other time in the past 600 years.

The Greenhouse Effect

As you have already learned, 50 percent of the radiation that enters the Earth's atmosphere is absorbed by the Earth's surface. This energy is then reradiated to the Earth's atmosphere as thermal energy. Gases, such as carbon dioxide and water vapor, can stop this energy from escaping into space by absorbing it and then radiating it back to the Earth. As a result, the Earth's atmosphere stays warm. This is similar to how a blanket keeps you warm at night. The Earth's heating process, in which the gases in the atmosphere trap thermal energy, is known as the **greenhouse effect.** This term is used because the Earth's atmosphere works much like a greenhouse, as shown in **Figure 10.**

Figure 10 *The gases in the atmosphere act like a layer of glass. The gases allow solar energy to pass through. But some of the gases trap thermal energy.*

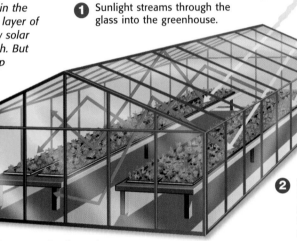

❶ Sunlight streams through the glass into the greenhouse.

❷ Sunlight is absorbed by objects inside the greenhouse. The objects radiate the energy as thermal energy.

❸ The glass stops the thermal energy from escaping to the outside.

Global Warming Not every gas in the atmosphere traps thermal energy. Those that do trap this energy are called *greenhouse gases*. In recent decades, many scientists have become concerned that an increase of these gases, particularly carbon dioxide, may be causing an increase in the greenhouse effect. These scientists have hypothesized that a rise in carbon dioxide as a result of human activity has led to increased global temperatures. A rise in average global temperatures is called **global warming.** If there were an increase in the greenhouse effect, global warming would result.

12

SCIENTISTS AT ODDS

While there is little argument about the accuracy of the greenhouse-effect model, there is much debate in the scientific community over whether the recent rise in global temperatures is due to global warming or a normal fluctuation in global temperatures. Scientists agree that the use of fossil fuels and CFCs as well as deforestation contribute to global warming in the greenhouse-effect model, but there is debate over which is the predominant cause. Have interested students learn more about global warming and stage a class debate over some of these issues.

The Radiation Balance For the Earth to remain livable, the amount of energy received from the sun and the amount of energy returned to space must be equal. As you saw in Figure 8, about 30 percent of the incoming energy is reflected back into space. Most of the 70 percent that is absorbed by the Earth and its atmosphere is also sent back into space. The balance between incoming energy and outgoing energy is known as the *radiation balance*. If greenhouse gases, such as carbon dioxide, continue to increase in the atmosphere, the radiation balance may be affected. Some of the energy that once escaped into space could be trapped. The Earth's temperatures would continue to rise, causing major changes in plant and animal communities.

Keeping the Earth Livable Some scientists argue that the Earth had warmer periods before humans ever walked the planet, so global warming may be a natural process. Nevertheless, many of the world's nations have signed a treaty to reduce activities that increase greenhouse gases in the atmosphere. Another step that is being taken to reduce high carbon dioxide levels in the atmosphere is the planting of millions of trees by volunteers, as shown in **Figure 11**.

Biology CONNECTION

Did you know that if you lived in Florida, your fingernails and toenails would grow faster than if you lived in Minnesota? Studies by scientists at Oxford University, in England, showed that warm weather helps tissue growth, while cold weather slightly slows it.

The study described in the Biology Connection on this page found that the average fingernail growth in the tropics is 1 mm a day, while in temperate regions, it is 0.8 mm a day. High altitude mountaineers are also familiar with this phenomenon. At altitudes above 6 km (19,600 feet), fingernail and hair growth slows dramatically due to the lack of oxygen. Cuts heal very slowly, and male mountaineers don't need to shave!

Figure 11 *Plants take in harmful carbon dioxide and give off oxygen, which we need to breathe.*

SECTION REVIEW

1. Describe three things that can happen to energy when it reaches the Earth's atmosphere.

2. How is energy transferred through the atmosphere?

3. What is the greenhouse effect?

4. **Inferring Relationships** How does the process of convection rely on conduction?

internet connect

SCLINKS
NSTA

TOPIC: Energy in the Atmosphere
GO TO: www.scilinks.org
*sci*LINKS **NUMBER:** HSTE360

13

4) Close

Quiz

1. What is radiation? (energy transferred as electromagnetic waves)

2. A metal spoon left in a bowl of hot soup feels hot. Which process—radiation, conduction, or convection—is mainly responsible for heating the spoon? (conduction)

3. What is a convection current? (the continual, circular movement of warm and cool particles in a liquid or gas)

4. How does a greenhouse stay warm? (Sunlight goes through the glass. Objects in the structure absorb some of the radiant energy. In turn, the objects radiate thermal energy. The glass prevents the energy from escaping, which warms the greenhouse.)

ALTERNATIVE ASSESSMENT

Writing Have students write a paragraph that compares and contrasts radiation, conduction, and convection. Ask students to explain how each process heats the atmosphere.

▼ *Answers to Section Review*

1. Answers will vary. Sample answer: Radiation can be absorbed by the Earth's surface. It can be absorbed by ozone, clouds, and the atmosphere or reflected by the Earth's surface and clouds.

2. Answers will vary. Sample answer: Energy is transferred through the atmosphere through radiation, conduction, and convection.

3. The greenhouse effect is the Earth's natural heating process by which gases in the atmosphere trap thermal energy.

4. Answers will vary. Sample answer: The air directly above the Earth's surface is heated by conduction. This warm air is then circulated through the atmosphere by convection currents.

Focus

Atmospheric Pressure and Winds

This section explains what wind is and describes how differences in atmospheric pressure cause air to move. Students will also learn about different wind patterns, including trade winds, jet streams, and local winds.

🔔 Bellringer

Tell students that **Figure 13** is an idealized model. Air doesn't actually follow the lines shown. Tell them that the Earth rotates west to east. Have students predict how this rotation would affect the movement of winds.

① Motivate

DEMONSTRATION

Air Movement This demonstration will show students how air moves from areas of high pressure to areas of low pressure. An area of high pressure can be created by filling a plastic container with ice. An area of low pressure can be created by heating a hot plate. Place the container of ice and the hot plate approximately 30 cm from each other. Make sure the container of ice is slightly higher than the hot plate so students can better observe the movement of air. Light a splint or long match, and let it burn for a few seconds. Extinguish the splint or match over the ice, and place the smoking end close to the ice. Observe the movement of the smoke. The smoke should move from the ice to the hot plate (from an area of high pressure to an area of low pressure).

Terms to Learn

wind	westerlies
Coriolis effect	polar easterlies
trade winds	jet streams

What You'll Do

- ◆ Explain the relationship between air pressure and wind direction.
- ◆ Describe the global patterns of wind.
- ◆ Explain the causes of local wind patterns.

Atmospheric Pressure and Winds

Sometimes it cools you. Other times it scatters tidy piles of newly swept trash. Still other times it uproots trees and flattens buildings, as shown in **Figure 12**. **Wind** is moving air. In this section you will learn about air movement and about the similarities and differences between different kinds of winds.

Figure 12 *In 1998, the winds from Hurricane Mitch reached speeds of 288 km/h, destroying entire towns in Honduras.*

Why Air Moves

Wind is created by differences in air pressure. The greater the pressure difference is, the faster the wind moves. This difference in air pressure is generally caused by the unequal heating of the Earth. For example, the air at the equator is warmer and less dense. This warm, less-dense air rises. As it rises it creates an area of low pressure. At the poles, however, the air is colder and more dense. Colder, more-dense air is heavier and sinks. This cold, sinking air creates areas of high pressure. Pressure differences in the atmosphere at the equator and at the poles cause air to move. Because air moves from areas of high pressure to areas of low pressure, winds generally move from the poles to the equator, as shown in **Figure 13**.

High pressure

Low pressure

High pressure

Figure 13 *Surface winds blow from polar high-pressure areas to equatorial low-pressure areas.*

14

Katabatic wind is the movement of air due to the influence of gravity. This flow can range from a gentle breeze to hurricane-force winds. The world's strongest katabatic winds occur in Antarctica because there is no shortage of cold air and the highest spot is near the center of the continent. Because the continent is roughly cone shaped, winds radiate from the South Pole, accelerating like a car rolling down a hill. Cold, dense air rushes down mountainsides, tumbles across the ice sheets, and spills out over the ocean. The winds can blow for months, and they sometimes reach speeds as great as 320 km/hr! Demonstrate this phenomenon using dry ice and a modeling-clay mountain.

Pressure Belts You may be imagining wind moving in one huge, circular pattern, from the poles to the equator. In fact, the pattern is much more complex. As warm air rises over the equator, it begins to cool. Eventually, it stops rising and moves toward the poles. At about 30° north and 30° south latitude, some of the cool air begins to sink. This cool, sinking air causes a high pressure belt near 30° north and 30° south latitude.

At the poles, cold air sinks. As this air moves away from the poles and along the Earth's surface, it begins to warm. As the air warms, the pressure drops, creating a low-pressure belt around 60° north and 60° south latitude. The circular patterns caused by the rising and sinking of air are called *convection cells,* as shown in **Figure 14.**

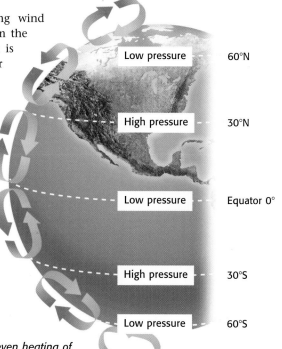

Low pressure — 60°N
High pressure — 30°N
Low pressure — Equator 0°
High pressure — 30°S
Low pressure — 60°S

Figure 14 *The uneven heating of the Earth produces pressure belts. These belts occur at about every 30° of latitude.*

Coriolis Effect Winds don't blow directly north or south. The movement of wind is affected by the rotation of the Earth. The Earth's rotation causes wind to travel in a curved path rather than in a straight line. The curving of moving objects, such as wind, by the Earth's rotation is called the **Coriolis effect.** Because of the Coriolis effect, the winds in the Northern Hemisphere curve to the right, and those in the Southern Hemisphere curve to the left.

To better understand how the Coriolis effect works, imagine rolling a marble across a Lazy Susan while it is spinning. What you might observe is shown in **Figure 15.**

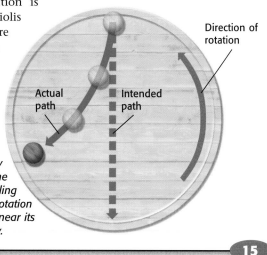

Direction of rotation

Actual path

Intended path

Figure 15 *Because of the Lazy Susan's rotation, the path of the marble curves instead of traveling in a straight line. The Earth's rotation affects objects traveling on or near its surface in much the same way.*

WEIRD SCIENCE

In addition to affecting ocean and atmospheric currents, the Coriolis effect can also be observed in river systems. Rivers in the Northern Hemisphere erode their right banks more than their left banks.

Because the Mississippi River and the Yukon River flow roughly north-south in sections, they are good examples of this effect.

Multicultural CONNECTION

Changes in atmospheric pressure are often said to affect fish. Egyptian fishermen notice that mullet move with the wind to prevent getting stuck in muddy water. According to Caribbean lore, a container of shark oil will grow cloudy when a hurricane is imminent. Have students find out about other organisms that might indicate changes in air pressure and other atmospheric phenomena.

MEETING INDIVIDUAL NEEDS

Learners Having Difficulty Try the following activity to help students who have problems understanding the Coriolis effect. You will need a globe, some flour, an eyedropper, red food coloring, and water. Mix a few drops of food coloring with water, and fill the eyedropper with the solution. Dust the globe thoroughly with flour. If the flour doesn't stick, mist the globe lightly with tap water and sprinkle the flour over the globe. Enlist a volunteer to slowly spin the globe counterclockwise to simulate Earth's rotation. Have another volunteer slowly drop water from the dropper from the top of the globe, at the North Pole. Students will observe that the water is deflected westward in the Northern Hemisphere. Have students demonstrate the Coriolis effect in the Southern Hemisphere by turning the globe upside down and rotating it counterclockwise.
Sheltered English

Directed Reading Worksheet Section 3

Section 3 • Atmospheric Pressure and Winds **15**

QuickLab

MATERIALS

- large clear-plastic container
- cold water
- packaging string (about 30 cm long)
- small plastic bottle with a narrow neck
- hot water
- red food coloring

Answers to QuickLab

6. The red warm water should rise. This activity models the circulation of air in the atmosphere. If the container were filled with cold water, the colored water would sink.

USING THE FIGURE

Have students use **Figure 16** to answer the following questions:

Where are the trade winds? (the winds that blow from 30° north and south latitudes to the equator)

Describe the motion of the trade winds in the Southern Hemisphere. (They move from the southeast to the northwest.)

How do the westerlies flow in the Northern Hemisphere? (The westerlies flow from the southwest to the northeast.)

What is the name of the windless zone that lies between the trade winds? (the doldrums)

QuickLab

Full of "Hot Air"

1. Fill a **large clear-plastic container** with **cold water.**
2. Tie the end of a **string** around the neck of a **small bottle.**
3. Fill the small bottle with **hot water,** and add a few drops of **red food coloring** until the water has changed color.
4. Without tipping the small bottle, lower it into the plastic container until it rests on the bottom.
5. Observe what happens.
6. What process does this activity model? What do you think will happen if you fill the small bottle with cold water instead? Try it!

TRY at HOME

Types of Winds

There are two main types of winds: local winds and global winds. Both types are caused by the uneven heating of the Earth's surface and by pressure differences. *Local winds* generally move short distances and can blow from any direction. *Global winds* are part of a pattern of air circulation that moves across the Earth. These winds travel longer distances than local winds, and they each travel in a specific direction. **Figure 16** shows the location and movement of major global wind systems. First let's review the different types of global winds, and later in this section we will discuss local winds.

Trade Winds In both hemispheres, the winds that blow from 30° latitude to the equator are called **trade winds.** The Coriolis effect causes the trade winds to curve, as shown in Figure 16. Early traders used the trade winds to sail from Europe to the Americas. This is how they became known as "trade winds."

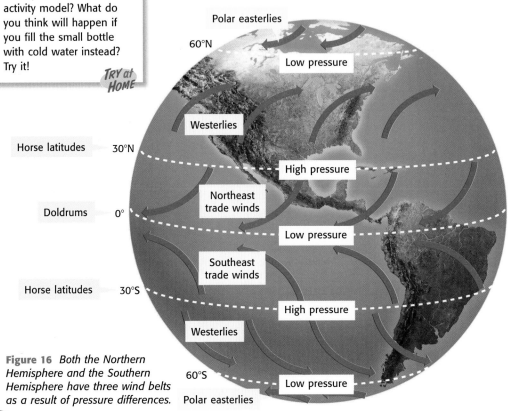

Figure 16 *Both the Northern Hemisphere and the Southern Hemisphere have three wind belts as a result of pressure differences.*

Science Bloopers

During a World War I naval engagement off the Falkland Islands, British gunners were astonished to see that their artillery shells were landing 100 yd to the left of German ships. The gunners had made corrections for the Coriolis effect at 50° north latitude, not 50° south of the equator. Consequently, their shells fell at a distance from the target equal to twice the Coriolis deflection!

The Doldrums and Horse Latitudes The trade winds of the Northern and Southern Hemispheres meet in an area of low pressure around the equator called the *doldrums*. In the doldrums there is very little wind because of the warm rising air. *Doldrums* comes from an Old English word meaning "foolish." Sailors were considered foolish if they got their ship stuck in these areas of little wind.

At about 30° north and 30° south latitude, sinking air creates an area of high pressure. This area is called the *horse latitudes*. Here the winds are weak. Legend has it that the name horse latitudes was given to these areas when sailing ships carried horses from Europe to the Americas. When the ships were stuck in this area due to lack of wind, horses were sometimes thrown overboard to save drinking water for the sailors.

Westerlies The **westerlies** are wind belts found in both the Northern and Southern Hemispheres between 30° and 60° latitude. The westerlies flow toward the poles in the opposite direction of the trade winds. The westerlies helped early traders return to Europe. Sailing ships, like the one in **Figure 17,** were designed to best use the wind to move the ship forward.

Figure 17 *This ship is a replica of Columbus's Santa Maria. If it had not sunk, the Santa Maria would have used the westerlies to return to Europe.*

Polar Easterlies The **polar easterlies** are wind belts that extend from the poles to 60° latitude in both hemispheres. The polar easterlies are formed from cold, sinking air moving from the poles toward 60° north and 60° south latitude.

Environment
CONNECTION

Humans have been using wind energy for thousands of years. Today wind energy is being tapped to produce electricity at wind farms. Wind farms are made up of hundreds of wind turbines that look like giant airplane propellers attached to towers. Together these wind turbines can produce enough electricity for an entire town.

To find out how to build a device that measures wind speed, turn to page 102 of the LabBook.

To find out how to build a device that measures wind speed, turn to page 102 of the LabBook.

17

The Earth's wind patterns are a rich biological oasis. Migrating birds use jet streams and local winds as aerial highways to reach their destinations. Insects and spiders also take advantage of wind currents— glider plane pilots have reported seeing air so thick with spiders that it looked like snow, and ships 800 km out at sea have been deluged with spiders falling from the sky! Dangling from the end of long silk streamers, spiders ride updrafts until they reach a wind current, occasionally reaching altitudes of 4 km. There they can travel for weeks, rolling in their streamers and dropping from the sky after covering up to 300 km. The spiders are a mobile meal for migrating purple martins. Students might enjoy finding radar entomology Web sites for people that track traveling insects.

MEETING INDIVIDUAL NEEDS

Advanced Learners Challenge interested students to explore the following question:

If we understand so much about the workings of the atmosphere, why can't we accurately predict weather from one week to the next?

Students will discover that scientists think the answer has to do with the chaotic nature of weather systems. Chaotic systems are so sensitive that slight variations at one point can result in huge changes later on. The behavior of chaotic systems is therefore very difficult to predict over more than a short period of time.

IS THAT A FACT!

Because the air descending over the horse latitudes has lost most of its moisture, the land around these latitudes receives very little precipitation. In fact, the Earth's largest deserts are in these areas.

Homework

Writing Have students watch or listen to a weather broadcast. Tell them to write down the meteorologist's comments about the local weather. Then have them write their interpretation of the forecast based on what they have learned about the atmosphere.

PORTFOLIO

LabBook **PG 102**
Go Fly a Bike!

GROUP ACTIVITY

Modeling Sea and Land Breezes Give each group two baking pans—one filled with sand and the other filled with ice. Groups should carefully warm the sand in an oven until it is very warm to the touch. Have the groups place the pans side-by-side. Then they should fold a cardboard windscreen in three places so that it wraps around both pans. As they hold a burning splint or stick of incense at the boundary between the pans, students should see smoke travel toward the hot sand in the same way that the wind blows toward the beach during the daytime. To simulate a land breeze, allow the sand to cool and replace the ice with warm water.

MATH and MORE

A pilot flying 950 km to Chicago is worried about a storm that will hit the city in 2 hours. The plane can fly at 500 km/h. A jet stream flowing in the opposite direction is moving at 250 km/h. If the plane must spend 10 minutes in the jet stream in order to climb above it, can the pilot make it to Chicago before the storm hits?

The pilot can beat the storm:

$(500 \text{ km/h} \times 1\frac{5}{6} \text{ h}) + (250 \text{ km/h} \times \frac{1}{6} \text{ h}) = 958 \text{ km in 2 hours}$

Teaching Transparency 166 "Sea and Land Breezes"

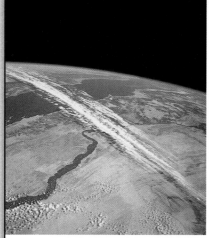
Figure 18 *The jet stream is the white stripe moving diagonally above the Earth.*

Jet Streams The **jet streams** are narrow belts of high-speed winds that blow in the upper troposphere and lower stratosphere, as shown in **Figure 18.** These winds often change speed and can reach maximum speeds of 500 km/h. Unlike other global winds, the jet streams do not follow regular paths around the Earth.

Knowing the position of the jet stream is important to both meteorologists and airline pilots. Because the jet stream controls the movement of storms, meteorologists can track a storm if they know the location of the jet stream. By flying in the direction of the jet stream, pilots can save time and fuel.

Local Winds Local winds are influenced by the geography of an area. An area's geography, such as a shoreline or a mountain, sometimes produces temperature differences that cause local winds like land and sea breezes, as shown in **Figure 19.** During the day, land heats up faster than water. The land heats the air above it. At night, land cools faster than water, cooling the air above the land.

Figure 19 Sea and Land Breezes

As warm air rises, it creates an area of low pressure over the land.

Warm air

The cool air moves toward the land, producing a *sea breeze.*

Cool air

Air over the water is cooler and creates an area of high pressure.

Air over land is cooler and creates an area of high pressure.

Cool air

The cool air moves toward the water, producing a *land breeze.*

Warm air

Air over the water is warmer and creates an area of low pressure.

Multicultural CONNECTION

The *chinook,* or "snow eater," is a dry wind that blows down the eastern side of the Rocky Mountains from New Mexico to Canada. Arapaho Indians gave the chinook its name because of its ability to melt large amounts of snow very quickly. Originating as moist air blowing off the Pacific Ocean, it heats up and loses moisture over the Rocky Mountains. When it reaches the Northwest, the chinook is warm and dry enough to melt a half meter of snow in a few hours! Have interested students research other local winds, such as the *sirocco,* the *Santa Ana,* or the *shamal,* that have a profound impact on people's lives.

Mountain and valley breezes are another example of local winds caused by an area's geography. Campers in mountain areas may feel a warm afternoon change into a cold night soon after the sun sets. The illustrations in **Figure 20** show you why.

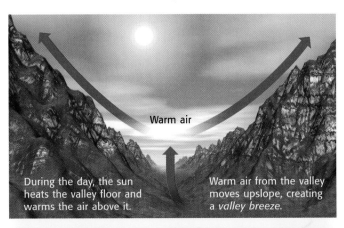

During the day, the sun heats the valley floor and warms the air above it.

Warm air

Warm air from the valley moves upslope, creating a *valley breeze*.

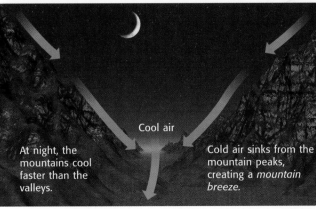

At night, the mountains cool faster than the valleys.

Cool air

Cold air sinks from the mountain peaks, creating a *mountain breeze*.

Figure 20 *During the day, a gentle breeze blows up the slopes. At night, cold air flows downslope and settles in the valley.*

SECTION REVIEW

1. How does the Coriolis effect affect wind movement?

2. What causes winds?

3. Compare and contrast global winds and local winds.

4. **Applying Concepts** Suppose you are vacationing at the beach. It is daytime and you want to go swimming in the ocean. You know the beach is near your hotel, but you don't know what direction it is in. How might the local wind help you find the ocean?

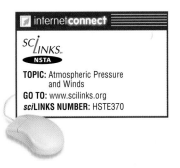

internet connect

*sci*LINKS_
NSTA

TOPIC: Atmospheric Pressure and Winds
GO TO: www.scilinks.org
*sci*LINKS NUMBER: HSTE370

Answers to MATHBREAK
500 km/h − 150 km/h = 350 km/h
3 h × 350 km/h = 1,050 km

4 Close

Quiz

1. What is wind? (air that flows between air masses of different pressures and temperatures)

2. How does air temperature over landmasses and adjacent bodies of water change between day and night? (During the day, the air is cooler over water. At night, the air is cooler over land.)

3. What is the Coriolis effect? (the deflection of moving objects due to Earth's rotation)

4. Compare and contrast the trade winds and the westerlies in the Northern Hemisphere. (Both are global wind systems that curve due to the Coriolis effect. Both result from differences in air pressure and temperature. The trade winds that lie between the equator and 30° north latitude blow from the northeast to the southwest. The westerlies lying between 30° and 60° north latitude blow from the southwest to the northeast.)

5. What are two kinds of breezes that result from local topography? (mountain and valley breezes)

ALTERNATIVE ASSESSMENT

Concept Mapping Have students create a concept map using the vocabulary and concepts in this section.

▼ **Answers to Section Review**

1. The Coriolis effect prevents winds from blowing directly north or south. Due to the Coriolis effect, trade winds in the Northern Hemisphere curve to the right and trade winds in the Southern Hemisphere curve to the left.

2. Winds are caused by the unequal heating of the Earth's surface and by pressure differences.

3. Local winds travel short distances and can blow from any direction. Global winds travel long distances and travel in specific directions.

4. During the day, a sea breeze is caused by the cooler air over the water moving toward the land. Walking toward the sea breeze would lead you to the ocean.

Focus

The Air We Breathe

This section defines and discusses air pollution. Students learn the difference between primary and secondary pollutants and explore sources of human-caused air pollution. Students then learn about some broader impacts of air pollution, such as acid precipitation and the ozone hole. Finally, students learn about the health effects of air pollution and what people can do to limit pollution.

Bellringer

Bring a filter mask to class. Have each student make a list of three situations in which one might wear such a mask. For example, surgeons wear such masks to prevent the transfer of disease-causing microbes, and sandblasters wear masks to avoid inhaling dust and paint chips. Tell students that some people living in areas with heavily polluted air wear such masks to protect themselves from impurities in the air they breathe.

1 Motivate

DISCUSSION

Explain that the air inside buildings may be polluted by a variety of sources. Ask students to list possible sources of indoor air pollution. If students have difficulty coming up with examples, tell them that air pollution is often invisible. Chalk dust, cooking oils, carpets, insulation, tobacco smoke, paints, glues, copier machines, space heaters, gas appliances, and fireplaces are just a few sources of indoor air pollution.

Terms to Learn

primary pollutants
secondary pollutants
acid precipitation

What You'll Do

◆ Describe the major types of air pollution.
◆ Name the major causes of air pollution.
◆ Explain how air pollution can affect human health.
◆ Explain how air pollution can be reduced.

The Air We Breathe

Air pollution, as shown in **Figure 21,** is not a new problem. By the middle of the 1700s, many of the world's large cities suffered from poor air quality. Most of the pollutants were released from factories and homes that burned coal for energy. Even 2,000 years ago, the Romans were complaining about the bad air in their cities. At that time the air was thick with the smoke from fires and the smell of open sewers. So you see, cities have always been troubled with air pollution. In this section you will learn about the different types of air pollution, their sources, and what the world is doing to reduce them.

Figure 21 *The air pollution in Mexico City is sometimes so dangerous that some people wear surgical masks when they go outside.*

Air Quality

Even "clean" air is not perfectly clean. It contains many pollutants from natural sources. These pollutants include dust, sea salt, volcanic gases and ash, smoke from forest fires, pollen, swamp gas, and many other materials. In fact, natural sources produce a greater amount of pollutants than humans do. But we have adapted to many of these natural pollutants.

Most of the air pollution mentioned in the news is a result of human activities. Pollutants caused by human activities can be solids, liquids, or gases. Human-caused air pollution, such as that shown in Figure 21, is most common in cities. As more people move to cities, urban air pollution increases.

 Directed Reading Worksheet Section 4

SCIENCE HUMOR

Q: What did the person say to the polluted air?

A: You take my breath away!

Types of Air Pollution

Air pollutants are generally described as either *primary pollutants* or *secondary pollutants*. **Primary pollutants** are pollutants that are put directly into the air by human or natural activity. **Figure 22** shows some examples of primary air pollutants.

Figure 22 *Exhaust from vehicles, ash from volcanic eruptions, and soot from smoke are all examples of primary pollutants.*

Secondary pollutants are pollutants that form from chemical reactions that occur when primary pollutants come in contact with other primary pollutants or with naturally occurring substances, such as water vapor. Many secondary pollutants are formed when a primary pollutant reacts with sunlight. Ozone and smog are examples of secondary pollutants. As you read at the beginning of this chapter, ozone is a gas in the stratosphere that is helpful and absorbs harmful rays from the sun. Near the ground, however, ozone is a dangerous pollutant that affects the health of all organisms. Ozone and smog are produced when sunlight reacts with automobile exhaust, as illustrated in **Figure 23.**

Figure 23 *Many large cities suffer from smog, especially those with a sunny climate and millions of automobiles.*

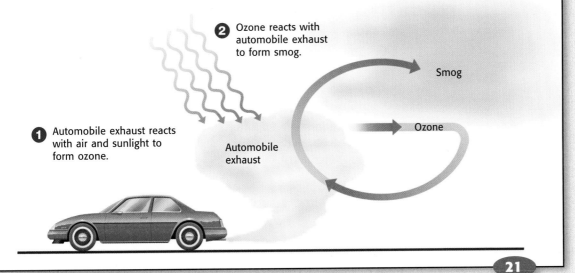

② Ozone reacts with automobile exhaust to form smog.

Smog

Ozone

① Automobile exhaust reacts with air and sunlight to form ozone.

Automobile exhaust

21

IS THAT A FACT!

In addition to forming as a reactant when certain pollutants are exposed to sunlight, ozone also forms during thunderstorms. Lightning provides the energy to change O_2 to O_3. In fact, the distinct smell people notice after an intense thunderstorm is ozone.

GUIDED PRACTICE

List the following pollutants on the board or on an overhead projector:

house dust, pollen, volcanic ash, soot, smog, ground-level ozone, acid rain

Beside the list, make a two-column table with the following column headings:

Primary pollutants, Secondary pollutants

Help students classify each pollutant as either a primary pollutant (house dust, pollen, volcanic ash, and soot) or a secondary pollutant (ground-level ozone, smog, and acid rain).

Sheltered English

MISCONCEPTION //// ALERT \\\\

Many people believe that polluted air must be visibly smoky or be brown or black in color. Stress that some of the most dangerous air pollutants are those that can't be seen with the naked eye. Challenge students to use the Internet to find out about the various pollutants monitored by the Environmental Protection Agency and other organizations that monitor air quality. Have students compile their results in a table that lists the acceptable amounts allowed in the air, the levels in your community, and the health problems associated with each pollutant.

Teaching Transparency 167
"The Formation of Smog"

Explain to students that much of the human-caused air pollution results from incomplete combustion. *Combustion,* another word for burning, is the process by which substances combine with oxygen rapidly, producing thermal energy. Byproducts are produced when a substance does not burn completely, as in an automobile engine. Many of these byproducts, such as carbon monoxide, are harmful to living organisms.

REAL-WORLD CONNECTION

Air quality varies greatly from place to place. Even in one location, air quality can change greatly from day to day. Have students research the air quality where they live. What are the sources of air pollution where you live? What are the weather conditions that lead to the worst and best air quality in your area?

RESEARCH

Writing Radon is a naturally occurring gas that results from the decay of uranium particularly in igneous rocks, such as granite. Have interested students research the air pollution and health problems associated with radon. Encourage students to assess the potential for significant radon concentrations in your community. Have students write a short informative essay based on their findings.

PORTFOLIO

Sources of Human-Caused Air Pollution

Human-caused air pollution comes from a variety of sources. The major source of air pollution today is transportation, as shown in **Figure 24.** Cars contribute about 60 percent of the human-caused air pollution in the United States. The oxides that come from car exhaust, such as nitrogen oxide, contribute to smog and acid rain. *Oxides* are chemical compounds that contain oxygen and other elements.

Figure 24 *Seventy percent of the carbon monoxide in the United States is produced by fuel-burning vehicles.*

Industrial Air Pollution Many industrial plants and electric power plants burn fossil fuels to get their energy. But burning fossil fuels causes large amounts of oxides to be released into the air, as shown in **Figure 25.** In fact, the burning of fossil fuels in industrial and electric power plants is responsible for 96 percent of the sulfur oxides released into the atmosphere.

Some industries also produce chemicals that form poisonous fumes. The chemicals used by oil refineries, chemical manufacturing plants, dry-cleaning businesses, furniture refinishers, and auto-body shops can add poisonous fumes to the air.

Figure 25 *This power plant burns coal to get its energy and releases sulfur oxides and particulates into the atmosphere.*

Indoor Air Pollution Air pollution is not limited to the outdoors. Sometimes the air inside a home or building is even worse than the air outside. The air inside a building can be polluted by the compounds found in household cleaners and cooking smoke. The compounds in new carpets, paints, and building materials can also add to indoor air pollution, especially if the windows and doors are tightly sealed to keep energy bills low.

22

Multicultural CONNECTION

Scientists have found high levels of airborne contaminants in the breast milk of Inuit women in Greenland and Arctic Canada. Researchers think the contaminants arrived in these remote areas by a process called global distillation. In this process, contaminants are redistributed around the globe by atmospheric currents. They tend to concentrate in northern areas for the same reason that water vapor condenses on cold glass: gaseous substances tend to condense at colder temperatures.

The Air Pollution Problem

Air pollution is both a local and global concern. As you have already learned, local air pollution, such as smog, generally affects large cities. Air pollution becomes a global concern when local pollution moves away from its source. Winds can move pollutants from one place to another, sometimes reducing the amount of pollution in the source area but increasing it in another place. For example, the prevailing winds carry air pollution created in the midwestern United States hundreds of miles to Canada. One such form of this pollution is acid precipitation.

Figure 26 *Acid precipitation can kill living things, such as fish and trees, by making their environment too acidic to live in.*

Acid Precipitation Precipitation that contains acids from air pollution is called **acid precipitation.** When fossil fuels are burned, they release oxides of sulfur and nitrogen into the atmosphere. When these oxides combine with water droplets in the atmosphere, they form sulfuric acid and nitric acid, which fall as precipitation. Acid precipitation has many negative effects on the environment, as shown in **Figure 26.**

The Ozone Hole Other global concerns brought about by air pollution include the warming of our planet and the ozone hole in the stratosphere. In the 1970s, scientists determined that some chemicals released into the atmosphere react with ozone in the ozone layer. The reaction results in a breakdown of ozone into oxygen, which does not block the sun's harmful ultraviolet rays. The loss of ozone creates an ozone hole, which allows more ultraviolet rays to reach the Earth's surface. **Figure 27** shows a satellite image of the ozone hole.

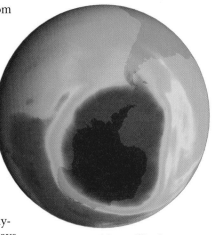

Figure 27 *This satellite image, taken in 1998, shows that the ozone hole, the dark blue area, is still growing.*

23

✷WEIRD SCIENCE

One reason the ozone layer is thinner over Antarctica involves a strange type of cloud. During winter in Antarctica, the stratosphere over the continent receives little light and temperatures can be below –80°C. In these conditions, chemicals in the air freeze and form *polar stratospheric clouds*. When light hits the clouds in spring, it catalyzes ozone-destroying reactions in cloud droplets, drastically reducing ozone concentrations over Antarctica.

GROUP ACTIVITY

Check the air quality near your school by collecting particulate matter. Remove the protective backing from an $8\frac{1}{2}$ × 11 in. sheet of clear contact paper, and place it over a sheet of graph paper. Pin the papers to a piece of cardboard with the sticky side up. Place the cardboard somewhere where it will be undisturbed for 1 day. Students can collect the contact paper and can use the grids on the graph paper and a magnifying glass to count the number of particles they collected. Students should also note particle sizes. As an extension, try this experiment at different times of the year and in different locations in your community.

INDEPENDENT PRACTICE

Writing Have interested students find out how the ozone hole has changed since it was first measured. Have students graph the values on both a yearly and seasonal basis and describe any trends they see. Have them compile their findings into a short report.

DEMONSTRATION

Demonstrate how acid rain affects limestone or marble. Put some limestone or marble chips into a beaker of vinegar. Let the chips sit a few days, and have students note any differences in the surface of the chips and in the acid solution. (Students should observe that the surface of the chips is pitted. The solution will be cloudy.)

3 Extend

RETEACHING

Have students try to answer the questions below without referring to their textbook.

How do primary air pollutants differ from secondary ones? (Primary pollutants enter the atmosphere from human activities and natural events. Secondary pollutants form when primary pollutants react with other primary pollutants or with naturally occurring substances in the air.)

How does smog form? (Smog forms when sunlight reacts with automobile exhaust to create ozone. Ozone then reacts with automobile exhaust to create smog.)

How does acid precipitation form? (Acid precipitation forms when fossil fuels are burned, releasing oxides of nitrogen and sulfur into the air. These oxides combine with moisture in the air to form acids that fall to Earth in rain, snow, sleet, and hail.)

CROSS-DISCIPLINARY FOCUS

Health Have students find out about respiratory diseases that can be aggravated by air pollution, such as asthma. Have students compile their findings into tables that list the diseases, their symptoms, how they are treated, the age groups most commonly afflicted, and the relationship between the diseases and air pollutants.

Answer to Activity

Answers will vary. Accept all reasonable responses.

BRAIN FOOD

Nonsmoking city dwellers are three to four times more likely to develop lung cancer than nonsmoking people in rural areas.

Effects on Human Health You step outside and notice a smoky haze. When you take a deep breath, your throat tingles and you begin to cough. Air pollution like this affects many cities around the world. For example, on March 17, 1992, in Mexico City, all children under the age of 14 were prohibited from going to school because of extremely high levels of air pollution. This is an extreme case, but daily exposure to small amounts of air pollution can cause serious health problems. Children, elderly people, and people with allergies, lung problems, and heart problems are especially vulnerable to the effects of air pollution. **Figure 28** illustrates some of the effects that air pollution has on the human body.

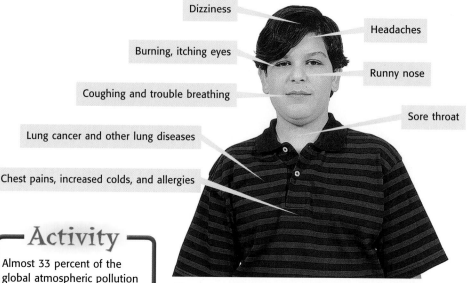

Dizziness
Headaches
Burning, itching eyes
Runny nose
Coughing and trouble breathing
Sore throat
Lung cancer and other lung diseases
Chest pains, increased colds, and allergies

Figure 28 *The Environmental Protection Agency blames air pollution for at least 2,000 new cases of cancer each year.*

Activity

Almost 33 percent of the global atmospheric pollution from carbon dioxide is caused by power plants that burn coal or other fossil fuels. We rely on these sources of power for a better way of life, but our use of them is polluting our air and worsening our quality of life. Use your school library or the Internet to find out about some other sources of electric power. What special problems does each source of energy bring with it?

TRY at HOME

24

Cleaning Up Our Act

Is all this talk about bad air making you a little choked up? Don't worry, help is on the way! In the United States, progress has been made in cleaning up the air. One reason for this progress is the Clean Air Act, which was passed by Congress in 1970. The Clean Air Act is a law that gives the Environmental Protection Agency (EPA) the authority to control the amount of air pollutants that can be released from any source, such as cars and factories. The EPA also checks air quality. If air quality worsens, the EPA can set stricter standards. What are car manufacturers and factories doing to improve air quality? Read on to find out.

internet connect

SCI LINKS
NSTA
TOPIC: Air Pollution
GO TO: www.scilinks.org
*sci*LINKS NUMBER: HSTE375

WEIRD SCIENCE

Ice core samples from Greenland show large-scale lead pollution in the atmosphere more than 2,000 years ago. The pollution can be traced to Roman silver mines in southern Spain. When smelting silver ore, large amounts of lead were released into the atmosphere.

Controlling Air Pollution from Vehicles The EPA has required car manufacturers to meet a certain standard for the exhaust that comes out of the tailpipe on cars. New cars now have devices that remove most of the pollutants from the car's exhaust as it exits the tailpipe. Car manufacturers are also making cars that run on fuels other than gasoline. Some of these cars run on hydrogen and natural gas, while others run on batteries powered by solar energy. The car shown in **Figure 29** is electric.

Are electric cars the cure for air pollution? Turn to page 33 and decide for yourself.

Figure 29 *Instead of having to refuel at a gas station, an electric car is plugged in to a recharging outlet.*

Controlling Air Pollution from Industry The Clean Air Act requires many industries to use scrubbers. A scrubber is a device that attaches to smokestacks to remove some of the more harmful pollutants before they are released into the air. One such scrubber is used in coal-burning power plants in the United States to remove ash and other particles from the smokestacks. Scrubbers prevent 22 million metric tons of ash from being released into the air each year.

Although we have a long way to go, we're taking steps in the right direction to keep the air clean for future generations.

SECTION REVIEW

1. How can the air inside a building be more polluted than the air outside?

2. Why might it be difficult to establish a direct link between air pollution and health problems?

3. How has the Clean Air Act helped to reduce air pollution?

4. **Applying Concepts** How is the water cycle affected by air pollution?

---

internetconnect

SCiLINKS.
NSTA

TOPIC: Air Pollution
GO TO: www.scilinks.org
*sci*LINKS **NUMBER:** HSTE375

25

---

4 Close

Quiz

1. Classify each of the following as either a primary or secondary air pollutant: smog, tobacco smoke, chalk dust, and acid rain. (Tobacco smoke and chalk dust are primary pollutants, while smog and acid rain are secondary pollutants.)

2. What are the three sources of outdoor air pollution? (motor vehicles, industries, electric power plants)

3. What are two health problems that can result from breathing polluted air? (dizziness, headaches, burning, itchy eyes, runny nose, coughing, shortness of breath, sore throat, lung cancer and other respiratory diseases, chest pain, colds, and allergies)

ALTERNATIVE ASSESSMENT

Have students use each of the following terms in a sentence that correctly conveys the meaning of the term:

scrubber, smog, acid precipitation, industrial pollutants, ozone hole, electric car, air quality

Critical Thinking Worksheet "The Extraordinary GBG5K"

Interactive Explorations CD-ROM "Moose Malady"

▼ *Answers to Section Review*

1. Answers will vary. Indoor air is polluted by household cleaners, air fresheners, smoke from cooking, as well as industrial compounds found in carpets, paints, building materials, and furniture.

2. Answers will vary. Accept all reasonable responses.

3. The Clean Air Act gives the EPA the authority to control the amount of air pollutants that can be released from any source. The EPA also monitors air quality; if the air quality worsens, the EPA can set stricter standards.

4. Answers will vary. Sample answer: Rainwater can become more acidic as a result of air pollution.

Discovery Lab

USING SCIENTIFIC METHODS

Under Pressure!
Teacher's Notes

Time Required

One 45-minute class period plus 15 minutes each day for 3–4 days

Lab Ratings

TEACHER PREP	▲▲
STUDENT SET-UP	▲▲▲
CONCEPT LEVEL	▲▲
CLEAN UP	▲▲

The materials listed on the student page are enough for a group of 2–4 students.

Safety Caution

Remind students to review all safety cautions and icons before beginning this lab activity.

Do not allow students to make a mercury barometer. Mercury fumes are dangerous.

Preparation Notes

A few weeks before the activity, collect daily weather newspaper clippings. A week before the activity, have students bring in large coffee cans. Jars can substitute for coffee cans in this experiment. For more accurate results, make sure students place their barometers in a shaded area.

Under Pressure!

You are planning a picnic with your friends, so you look in the newspaper for the weather forecast. The temperature this afternoon should be in the low 80s. This temperature sounds quite comfortable! But you notice that the newspaper's forecast also includes the barometer reading. What's a barometer? And what does the reading tell you? In this activity, you will build your own barometer and will discover what this tool can tell you.

MATERIALS

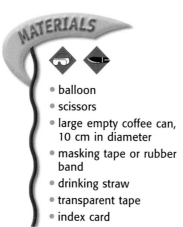

- balloon
- scissors
- large empty coffee can, 10 cm in diameter
- masking tape or rubber band
- drinking straw
- transparent tape
- index card

Ask a Question

1 How can I make a tool that measures changes in atmospheric pressure?

Form a Hypothesis

2 In your ScienceLog, write a few sentences that answer the question above.

Conduct an Experiment

3 Stretch and blow up the balloon. Then let the air out. This step will let your barometer be more sensitive to changes in atmospheric pressure.

4 Cut off the end of the balloon that you put in your mouth to blow it up. Stretch the balloon over the mouth of the coffee can. Attach the balloon to the can with the tape or the rubber band.

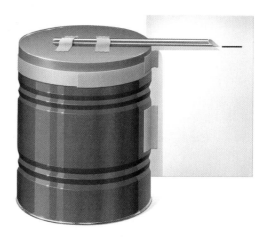

26

Datasheets for LabBook

Science Skills Worksheet
"Using Your Senses"

Terry J. Rakes
Elmwood Jr. High
Rogers, Arkansas

5 Cut one end of the straw at an angle to make a pointer.

6 Place the straw with the pointer pointed away from the center of the stretched balloon. Look at the illustration below. Place the straw so that 5 cm of the end of the straw hang over the edge of the can. Tape the straw to the balloon.

7 Tape the index card to the can near the straw. Congratulations! You have just made a barometer!

8 Place the barometer outside for 3–4 days. On each day, mark on the index card where the straw points.

Analyze the Results

9 What factors affect how your barometer works? Explain your answer.

10 What does it mean when the straw moves up?

11 What does it mean when the straw moves down?

Draw Conclusions

12 Compare your results with the barometric pressures listed in your local newspaper. What kind of weather is associated with high pressure? What kind of weather is associated with low pressure?

13 Does the barometer you built support your hypothesis? Explain your answer.

Going Further
Now you can calibrate your barometer! Get the weather section from your local newspaper for the same three or four days that you are testing your barometer. Find the barometer reading in the newspaper for each day, and record it beside that day's mark on your index card. Divide the markings on the index card into equal spaces. Write the matching barometric pressures on the card.

Answers

9. Temperature and pressure will affect how the barometer works. At higher temperatures, air pressure decreases, relieving pressure on the balloon and making the straw point downward. At lower temperatures, the air pressure increases, increasing pressure on the balloon and making the straw point upward.

10. An upward movement of the straw means that the atmospheric pressure is increasing. Pressure is pushing on top of the balloon, causing the pointer to rise.

11. A downward movement of the straw means that the atmospheric pressure is decreasing.

12. Clear, dry days are associated with high pressure. Cloudy, rainy, or humid days are associated with low pressure. A sudden drop in air pressure usually indicates that a storm is on the way.

13. Answers will vary.

Chapter Highlights

Chapter Highlights

VOCABULARY DEFINITIONS

SECTION 1

atmosphere a mixture of gases that surrounds a planet, such as Earth

air pressure the measure of the force with which air molecules push on a surface

altitude the height of an object above the Earth's surface

troposphere the lowest layer of the atmosphere

stratosphere the atmospheric layer above the troposphere

ozone a gas molecule that is made up of three oxygen atoms and that absorbs ultraviolet radiation from the sun

mesosphere the coldest layer of the atmosphere

thermosphere the uppermost layer of the atmosphere

SECTION 2

radiation the transfer of energy as electromagnetic waves, such as visible light or infrared waves

conduction the transfer of thermal energy from one material to another by direct contact; conduction can also occur within a substance

convection the transfer of thermal energy by the circulation or movement of a liquid or gas

greenhouse effect the natural heating process of a planet, such as the Earth, by which gases in the atmosphere trap thermal energy

global warming a rise in average global temperatures

SECTION 1

Vocabulary

atmosphere *(p. 4)*
air pressure *(p. 5)*
altitude *(p. 5)*
troposphere *(p. 7)*
stratosphere *(p. 7)*
ozone *(p. 7)*
mesosphere *(p. 8)*
thermosphere *(p. 8)*

Section Notes

- The atmosphere is a mixture of gases.
- Nitrogen and oxygen are the two most abundant atmospheric gases.
- Throughout the atmosphere, there are changes in air pressure, temperature, and gases.
- Air pressure decreases as altitude increases.

- Temperature differences in the atmosphere are a result of the way solar energy is absorbed as it moves downward through the atmosphere.
- The troposphere is the lowest and densest layer of the atmosphere. All weather occurs in the troposphere.
- The stratosphere contains the ozone layer, which protects us from harmful radiation.
- The mesosphere is the coldest layer of the atmosphere.
- The uppermost atmospheric layer is the thermosphere.

SECTION 2

Vocabulary

radiation *(p. 10)*
conduction *(p. 11)*
convection *(p. 11)*
greenhouse effect *(p. 12)*
global warming *(p. 12)*

Section Notes

- The Earth receives energy from the sun by radiation.
- Energy that reaches the Earth's surface is absorbed or reflected.
- Energy is transferred through the atmosphere by conduction and convection.
- The greenhouse effect is caused by gases in the atmosphere that trap thermal energy reflected off and radiated from the Earth's surface.

Labs

Boiling Over! *(p. 100)*

☑ Skills Check

Math Concepts

FLYING AGAINST THE JET STREAM The groundspeed of an airplane can be affected by the jet stream. The jet stream can push an airplane toward its final destination or slow it down. To find the groundspeed of an airplane, you either add or subtract the wind speed, depending on whether the airplane is moving with or against the jet stream. For example, if an airplane is traveling at an airspeed of 400 km/h and is moving with a 100 km/h jet stream, you would add the jet stream speed to the airspeed of the airplane to calculate the groundspeed.

$$400 \text{ km/h} + 100 \text{ km/h} = 500 \text{ km/h}$$

To calculate the groundspeed of an airplane traveling at 400 km/h that is moving into a 100 km/h jet stream, you would subtract the jet-stream speed from the airspeed of the airplane.

$$400 \text{ km/h} - 100 \text{ km/h} = 300 \text{ km/h}$$

Visual Understanding

GLOBAL WINDS Study Figure 16 on page 16 to review the global wind belts that result from air pressure differences.

Lab and Activity Highlights

Under Pressure! PG 26

Boiling Over! PG 100

Go Fly a Bike! PG 102

Datasheets for LabBook (blackline masters for these labs)

SECTION 3

Vocabulary

wind *(p. 14)*

Coriolis effect *(p. 15)*

trade winds *(p. 16)*

westerlies *(p. 17)*

polar easterlies *(p. 17)*

jet streams *(p. 18)*

Section Notes

• At the Earth's surface, winds blow from areas of high pressure to areas of low pressure.

• Pressure belts exist approximately every 30° of latitude.

• The Coriolis effect makes wind curve as it moves across the Earth's surface.

• Global winds are part of a pattern of air circulation across the Earth and include the trade winds, the westerlies, and the polar easterlies.

• Local winds move short distances, can blow in any direction, and are influenced by geography.

Labs

Go Fly a Bike! *(p. 102)*

SECTION 4

Vocabulary

primary pollutants *(p. 21)*

secondary pollutants *(p. 21)*

acid precipitation *(p. 23)*

Section Notes

• Air pollutants are generally classified as primary or secondary pollutants.

• Human-caused pollution comes from a variety of sources, including factories, cars, and homes.

• Air pollution can heighten problems associated with allergies, lung problems, and heart problems.

• The Clean Air Act has reduced air pollution by controlling the amount of pollutants that can be released from cars and factories.

SECTION 3

wind moving air

Coriolis effect the curving of moving objects from a straight path due to the Earth's rotation

trade winds the winds that blow from 30° latitude to the equator

westerlies wind belts found in both the Northern and Southern Hemispheres between 30° and 60° latitude

polar easterlies wind belts that extend from the poles to 60° latitude in both hemispheres

jet streams narrow belts of high-speed winds that blow in the upper troposphere and the lower stratosphere

SECTION 4

primary pollutants pollutants that are put directly into the air by human or natural activity

secondary pollutants pollutants that form from chemical reactions that occur when primary pollutants come in contact with other primary pollutants or with naturally occurring substances, such as water vapor

acid precipitation precipitation that contains acids due to air pollution

---

 internet**connect**

 GO TO: go.hrw.com

Visit the **HRW** Web site for a variety of learning tools related to this chapter. Just type in the keyword:

KEYWORD: HSTATM

  SCI**LINKS**sm

N S T A

GO TO: www.scilinks.org

Visit the **National Science Teachers Association** on-line Web site for Internet resources related to this chapter. Just type in the *sci*LINKS number for more information about the topic:

TOPIC: Composition of the Atmosphere	*sci*LINKS NUMBER: HSTE355
TOPIC: Energy in the Atmosphere	*sci*LINKS NUMBER: HSTE360
TOPIC: The Greenhouse Effect	*sci*LINKS NUMBER: HSTE365
TOPIC: Atmospheric Pressure and Winds	*sci*LINKS NUMBER: HSTE370
TOPIC: Air Pollution	*sci*LINKS NUMBER: HSTE375

29

 Vocabulary Review Worksheet

Blackline masters of these Chapter Highlights can be found in the **Study Guide.**

Lab and Activity Highlights

LabBank

 Whiz-Bang Demonstrations, Blue Sky

EcoLabs & Field Activities, That Greenhouse Effect!

Calculator-Based Labs, The Greenhouse Effect

Long-Term Projects & Research Ideas, A Breath of Fresh Ether?

Interactive Explorations CD-ROM

 CD 2, Exploration 3, "Moose Malady"

USING VOCABULARY

1. Air pressure is the measure of the force with which air molecules are pushing on a surface. Altitude is the height of an object above the Earth's surface.

2. The troposphere is the lowest layer of the Earth's atmosphere. The thermosphere is the uppermost atmospheric layer.

3. The greenhouse effect is the Earth's natural heating process, by which gases in the atmosphere trap thermal energy. Global warming is a rise in average global temperatures possibly due to an increase in the greenhouse effect.

4. Convection is the transfer of thermal energy by the circulation of a liquid or gas. Conduction is the transfer of energy from one material to another by direct contact.

5. Global winds are a part of a pattern of air circulation that moves across the Earth. Local winds generally move short distances and can blow from any direction.

6. Primary pollutants are pollutants that are put directly into the air by human or natural activity. Secondary pollutants form from chemical reactions that occur when primary pollutants come in contact with other primary pollutants or with naturally occurring substances.

UNDERSTANDING CONCEPTS

Multiple Choice

7. b
8. c
9. b
10. c
11. a
12. b
13. d
14. a
15. b
16. d

Chapter Review

USING VOCABULARY

Explain the difference between the following sets of words:

1. air pressure/altitude

2. troposphere/thermosphere

3. greenhouse effect/global warming

4. convection/conduction

5. global wind/local wind

6. primary pollutant/secondary pollutant

UNDERSTANDING CONCEPTS

Multiple Choice

7. What is the most abundant gas in the air that we breathe?
   a. oxygen
   b. nitrogen
   c. hydrogen
   d. carbon dioxide

8. The major source of oxygen for the Earth's atmosphere is
   a. sea water.
   b. the sun.
   c. plants.
   d. animals.

9. The bottom layer of the atmosphere, where almost all weather occurs, is the
   a. stratosphere.
   b. troposphere.
   c. thermosphere.
   d. mesosphere.

10. About ___?___ percent of the solar energy that reaches the outer atmosphere is absorbed at the Earth's surface.
    a. 20
    b. 30
    c. 50
    d. 70

11. The ozone layer is located in the
    a. stratosphere.
    b. troposphere.
    c. thermosphere.
    d. mesosphere.

12. How does most thermal energy in the atmosphere move?
    a. conduction
    b. convection
    c. advection
    d. radiation

13. The balance between incoming and outgoing energy is called ___?___.
    a. convection
    b. conduction
    c. greenhouse effect
    d. radiation balance

14. Most of the United States is located in which prevailing wind belt?
    a. westerlies
    b. northeast trade winds
    c. southeast trade winds
    d. doldrums

15. Which of the following is not a primary pollutant?
    a. car exhaust
    b. acid precipitation
    c. smoke from a factory
    d. fumes from burning plastic

30

16. The Clean Air Act
    a. controls the amount of air pollutants that can be released from most sources.
    b. requires cars to run on fuels other than gasoline.
    c. requires many industries to use scrubbers.
    d. (a) and (c) only

Short Answer

17. Why does the atmosphere become less dense as altitude increases?

18. Explain why air rises when it is heated.

19. What causes temperature changes in the atmosphere?

20. What are secondary pollutants, and how are they formed? Give an example.

Concept Mapping

21. Use the following terms to create a concept map: altitude, air pressure, temperature, atmosphere.

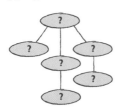

CRITICAL THINKING AND PROBLEM SOLVING

Write one or two sentences to answer the following questions:

22. What is the relationship between the greenhouse effect and global warming?

23. How do you think the Coriolis effect would change if the Earth were to rotate twice as fast? Explain.

24. Without the atmosphere, the Earth's surface would be very different. What are several ways that the atmosphere affects the Earth?

MATH IN SCIENCE

25. Wind speed is measured in miles per hour and in knots. One mile (statute mile or land mile) is 5,280 ft. One nautical mile (or sea mile) is 6,076 ft. Speed in nautical miles is measured in knots. Calculate the wind speed in knots if the wind is blowing at 25 mi/h.

INTERPRETING GRAPHICS

Use the wind-chill chart to answer the questions below.

Wind-Chill Chart

Wind Speed		Actual thermometer reading (°F)				
		40	30	20	10	0
Knots	mph	Equivalent temperature (°F)				
Calm		40	30	20	10	0
4	5	37	27	16	6	−5
9	10	28	16	4	−9	−21
13	15	22	9	−5	−18	−36
17	20	18	4	−10	−25	−39
22	25	16	0	−15	−29	−44
26	30	13	−2	−18	−33	−48
30	35	11	−4	−20	−35	−49

26. If the wind speed is 20 mi/h and the temperature is 40°F, how cold will the air seem?

27. If the wind speed is 30 mi/h and the temperature is 20°F, how cold will the air seem?

Reading Check-up

Take a minute to review your answers to the Pre-Reading Questions found at the bottom of page 2. Have your answers changed? If necessary, revise your answers based on what you have learned since you began this chapter.

31

Short Answer

17. As altitude increases, there are fewer gas molecules. Gravity pulls much of the atmosphere's gas molecules close to the Earth's surface.

18. Air rises as it is heated because it becomes less dense.

19. The temperature differences in the atmosphere result mainly from the way solar energy is absorbed as it moves downward through the atmosphere. Some layers are warmer because they contain gases that absorb solar energy.

20. Secondary pollutants form when a primary pollutant reacts with other primary pollutants or with naturally occurring substances. Smog and ozone are examples of secondary pollutants.

Concept Mapping

21. An answer to this exercise can be found in the front of this book.

CRITICAL THINKING AND PROBLEM SOLVING

22. An increase in the greenhouse effect could result in global warming.

23. The Coriolis effect would be more pronounced if the Earth rotated twice as fast. Winds are affected by the rotation of the Earth; if the speed were increased, the curvature would be more pronounced.

24. Answers will vary. Sample answer: The atmosphere protects living organisms from harmful radiation from the sun. Without the atmosphere, more of this radiation would reach the Earth's surface.

MATH IN SCIENCE

25. 22 knots

INTERPRETING GRAPHICS

26. 18°F
27. −18°F

Background

Particulate matter is one of the major forms of air pollution, along with carbon monoxide (a toxic gas), sulfur dioxides (which contribute to acid rain and human respiratory problems), and volatile organic compounds, or VOCs (organic chemicals that vaporize and produce toxic fumes).

Particulates are often formed during mechanical processes that break down materials. These include blasting, drilling, and grinding. Some organic matter, such as certain bacteria, are also considered particulates.

The growing problem of particulate pollution has been addressed by the federal government as well as by many local governments and communities. The Clean Air Act, passed by Congress in 1970, sets maximum emission levels for automobiles and industrial sources of pollution. New filtering technology has also helped industries to reduce the amount of particulates being released into the atmosphere.

- In most cases, the majority of the particulates found indoors in dust are particles of human skin.
- Certain types of asbestos are particularly dangerous as particulates. When inhaled, these asbestos fibers scar the lungs, inhibiting breathing and eventually causing cancer.

Health WATCH

Particles in the Air

Take a deep breath. You have probably just inhaled thousands of tiny specks of dust, pollen, and other particles. These particles, called particulates, are harmless under normal conditions. But if concentrations of particulates get too high or if they consist of harmful materials, they are considered to be a type of air pollution.

Because many particulates are very small, our bodies' natural filters, such as nasal hairs and mucous membranes, cannot filter all of them out. When inhaled, particulates can cause irritation in the lungs. Over time, this irritation can lead to diseases such as bronchitis, asthma, and emphysema. The danger increases as the level of particulates in the air increases.

▲ *When the ash from Mount St. Helens settled from the air, it created scenes like this one.*

Where There's Smoke . . .

Unfortunately, dust and pollen are not the only forms of particulates. Many of the particulates in the air come from the burning of various materials. For example, when wood is burned, it releases particles of smoke, soot, and ash into the air. Some of these are so small that they can float in the air for days. The burning of fuels such as coal, oil, and gasoline also creates particulates. The particulates from these sources can be very dangerous in high concentrations. That's why particulate concentrations are one measure of air quality. Large concentrations of particulates are visible in the air. Along with other pollutants, particulates are

what make polluted air look brown or yellowish brown. But don't be fooled—even air that appears clean can be polluted.

Eruptions of Particulates

Volcanoes can be the source of incredible amounts of particulates. For example, when Mount St. Helens erupted in 1980, it launched thousands of tons of ash into the surrounding air. The air was so thick with ash that the area became as dark as night. For several hours, the ash completely blocked the light from the sun. When the ash finally settled from the air, it covered the surrounding landscape like a thick blanket of snow. This layer of ash killed plants and livestock for several kilometers around the volcano.

One theory to explain the extinction of dinosaurs is that a gargantuan meteorite hit the Earth with such velocity that the resulting impact created enough dust to block out the sun for years. During this dark period, plants were unable to grow and therefore could not support the normal food chains. Consequently, the dinosaurs died out.

Do Filters Really Filter?

▶ Since the burning of most substances creates particulates, there must be particulates in cigarette smoke. Do some research to find out if the filters on cigarettes are effective at preventing particulates from entering the smoker's body. Your findings may surprise you!

32

Answers to Do Filters Really Filter?

Answers will vary. Cigarette filters absorb some of the particulates found in cigarette smoke but not all of them.

SCIENTIFIC DEBATE
A Cure for Air Pollution?

A Cure for Air Pollution?

Automobile emissions are responsible for at least half of all urban air pollution and a quarter of all carbon dioxide released into the atmosphere. Therefore, the production of a car that emits no polluting gases in its exhaust is a significant accomplishment. The only such vehicle currently available is the electric car. Electric cars are powered by batteries, so they do not produce exhaust gases. Supporters believe that switching to electric cars will reduce air pollution in this country. But critics believe that taxpayers will pay an unfair share for this switch and that the reduction in pollution won't be as great as promised.

▲ *Will a switch to electric cars such as this one reduce air pollution?*

Electric Cars Will Reduce Air Pollution

Even the cleanest and most modern cars emit pollutants into the air. Supporters of a switch to electric cars believe the switch will reduce pollution in congested cities. But some critics suggest that a switch to electric cars will simply move the source of pollution from a car's tailpipe to the power plant's smokestack. This is because most electricity is generated by burning coal.

In California, electric cars would have the greatest impact. Here most electricity is produced by burning natural gas, which releases less air pollution than burning coal.

Nuclear plants and dams release no pollutants in the air when they generate electricity. Solar power and wind power are also emission-free ways to generate electricity. Supporters argue that a switch to electric cars will reduce air pollution immediately and that a further reduction will occur when power plants convert to these cleaner sources of energy.

Electric Cars Won't Solve the Problem

Electric cars are inconvenient because the batteries have to be recharged so often. The batteries also have to be replaced every 2 to 3 years. The nation's landfills are already crowded with conventional car batteries, which contain acid and metals that may pollute ground water. A switch to electric cars would aggravate this pollution problem

because the batteries have to be replaced so often.

Also, electric cars will likely replace the cleanest cars on the road, not the dirtiest. A new car may emit only one-tenth of the pollution emitted by an older model. If an older car's pollution-control equipment does not work properly, it may emit 100 times more pollution than a new car. But people who drive older, poorly maintained cars probably won't be able to afford expensive electric cars. Therefore, the worst offenders will stay on the road, continuing to pollute the air.

Analyze the Issue

▶ Do you think electric cars are the best solution to the air pollution problem? Why or why not? What are some alternative solutions for reducing air pollution?

33

Background

Electric vehicles have been around for more than 150 years. Their first commercial use was in 1897, when New York City established a fleet of electric taxis. In the years 1899 and 1900 electric vehicles in America outsold all other types of cars. The 1902 Wood's Phaeton had a range of 29 km (18 mi), a top speed of 23 km/h (14 mph), and a price of $2,000.

Much has changed since the time of the Phaeton. Cars powered by hydrogen-oxygen fuel cells are classified as negative-emission vehicles. The cars use oxygen from the air to react with hydrogen in the fuel cell. During the process, harmful particles are removed from the air leaving it cleaner than it was before.

Answers to Analyze the Issue

Answers will vary. Alternative solutions for reducing air pollution might include educating people about the seriousness of air pollution problems and providing incentives for people to use mass transportation. Other ideas might include alternative fuel sources and more-efficient use of the resources we have. Students might also note that slowing the rate of deforestation would help reduce air pollution significantly.

Chapter Organizer

CHAPTER ORGANIZATION	TIME MINUTES	OBJECTIVES	LABS, INVESTIGATIONS, AND DEMONSTRATIONS
Chapter Opener pp. 34–35	45	National Standards: UCP 2, SAI 1, SPSP 3, 4, ES 1i, 1j	**Start-Up Activity,** A Meeting of the Masses, p. 35
Section 1 Water in the Air	120	▶ Explain how water moves through the water cycle. ▶ Define *relative humidity*. ▶ Explain what the dew point is and its relation to condensation. ▶ Describe the three major cloud forms. ▶ Describe the four major types of precipitation. UCP 2, 3, SAI 1, SPSP 3, ES 1f, 1i; Labs UCP 3, SAI 1	**QuickLab,** Out of Thin Air, p. 39 **Demonstration,** Hail Formation, p. 42 in ATE **Discovery Lab,** Let It Snow! p. 107 **Datasheets for LabBook,** Let It Snow! **Whiz-Bang Demonstrations,** It's Raining Again **Calculator-Based Labs,** Relative Humidity
Section 2 Air Masses and Fronts	90	▶ Explain how air masses are characterized. ▶ Describe the four major types of air masses that influence weather in the United States. ▶ Describe the four major types of fronts. ▶ Relate fronts to weather changes. HNS 3, ES 1j	**Demonstration,** p. 44 in ATE **Whiz-Bang Demonstrations,** When Air Bags Collide **Long-Term Projects & Research Ideas,** A Storm on the Horizon
Section 3 Severe Weather	90	▶ Explain what lightning is. ▶ Describe the formation of thunderstorms, tornadoes, and hurricanes. ▶ Describe the characteristics of thunderstorms, tornadoes, and hurricanes. SPSP 3, 4, ES 1i, 1j	**Demonstration,** p. 48 in ATE **Inquiry Labs,** When Disaster Strikes
Section 4 Forecasting the Weather	90	▶ Describe the different types of instruments used to take weather measurements. ▶ Explain how to interpret a weather map. ▶ Explain why weather maps are useful. UCP 3, ST 2, SPSP 5, HNS 1; Labs SAI 1, ST 1	**Demonstration,** p. 54 in ATE **Skill Builder,** Watching the Weather, p. 104 **Datasheets for LabBook,** Watching the Weather **Making Models,** Gone with the Wind, p. 58 **Datasheets for LabBook,** Gone with the Wind **EcoLabs & Field Activities,** Rain Maker or Rain Faker?

See page **T23** *for a complete correlation of this book with the*

NATIONAL SCIENCE EDUCATION STANDARDS.

TECHNOLOGY RESOURCES

 Guided Reading Audio CD
English or Spanish, Chapter 2

 One-Stop Planner CD-ROM with Test Generator

 CNN. Eye on the Environment, Hazy Days, Segment 1

 Science Discovery Videodiscs
Image and Activity Bank with Lesson Plans: Tracking Tornadoes, Tracking a Hurricane

CLASSROOM WORKSHEETS, TRANSPARENCIES, AND RESOURCES	SCIENCE INTEGRATION AND CONNECTIONS	REVIEW AND ASSESSMENT
Directed Reading Worksheet **Science Puzzlers, Twisters & Teasers**		
Directed Reading Worksheet, Section 1 **Transparency 168,** The Water Cycle **Transparency 169,** Cloud Types Based on Form and Altitude	**Connect to Physical Science,** p. 36 in ATE **MathBreak,** Relating Relative Humidity, p. 37 **Connect to Life Science,** p. 37 in ATE **Multicultural Connection,** pp. 38, 39, 42 in ATE **Math and More,** p. 40 in ATE **Cross-Disciplinary Focus,** p. 41 in ATE **Connect to Physical Science,** p. 42 in ATE	**Self-Check,** p. 38 **Section Review,** p. 39 **Homework,** pp. 40, 41 in ATE **Section Review,** p. 43 **Quiz,** p. 43 in ATE **Alternative Assessment,** p. 43 in ATE
Directed Reading Worksheet, Section 2 **Transparency 170,** Air Masses in North America **Transparency 171,** Cold and Warm Fronts **Transparency 172,** Occluded and Stationary Fronts	**Multicultural Connection,** p. 45 in ATE **Connect to Life Science,** p. 45 in ATE **Cross-Disciplinary Focus,** p. 46 in ATE	**Section Review,** p. 47 **Quiz,** p. 47 in ATE **Alternative Assessment,** p. 47 in ATE
Transparency 268, How Lightning Forms **Directed Reading Worksheet,** Section 3 **Transparency 173,** How a Tornado Forms **Transparency 174,** A Cross Section of a Hurricane **Reinforcement Worksheet,** Precipitation Situations	**Physics Connection,** p. 49 **Math and More,** p. 49 in ATE **Real-World Connection,** p. 49 in ATE **Connect to Physical Science,** p. 49 in ATE **Cross-Disciplinary Focus,** p. 51 in ATE **Astronomy Connection,** p. 53 **Holt Anthology of Science Fiction,** *All Summer in a Day*	**Homework,** p. 51 in ATE **Section Review,** p. 53 **Quiz,** p. 53 in ATE **Alternative Assessment,** p. 53 in ATE
Directed Reading Worksheet, Section 4 **Math Skills for Science Worksheet,** Using Temperature Scales **Critical Thinking Worksheet,** Commanding the Sky	**Real-World Connection,** p. 55 in ATE **Connect to Life Science,** p. 55 in ATE **Math and More,** p. 56 in ATE **Cross-Disciplinary Focus,** p. 56 in ATE **Careers:** Meteorologist—Cristy Mitchell, p. 64	**Section Review,** p. 57 **Quiz,** p. 57 in ATE **Alternative Assessment,** p. 57 in ATE

 internetconnect

 Holt, Rinehart and Winston On-line Resources
go.hrw.com

For worksheets and other teaching aids related to this chapter, visit the HRW Web site and type in the keyword: **HSTWEA**

 National Science Teachers Association
www.scilinks.org

Encourage students to use the *sci*LINKS numbers listed in the internet connect boxes to access information and resources on the **NSTA** Web site.

END-OF-CHAPTER REVIEW AND ASSESSMENT

Chapter Review in Study Guide
Vocabulary and Notes in Study Guide
Chapter Tests with Performance-Based Assessment, Chapter 2 Test, Performance-Based Assessment 2
Concept Mapping Transparency 16

Chapter Resources & Worksheets

Visual Resources

TEACHING TRANSPARENCIES

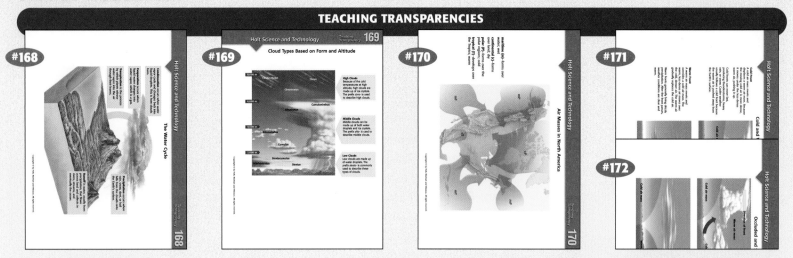

#168 — Holt Science and Technology — The Water Cycle — 168

#169 — Holt Science and Technology — Teaching Transparency 169 — Cloud Types Based on Form and Altitude

#170 — Holt Science and Technology — Air Masses in North America — 170

#171 — Holt Science and Technology — Cold and

#172 — Holt Science and Technology — Occluded and

TEACHING TRANSPARENCIES

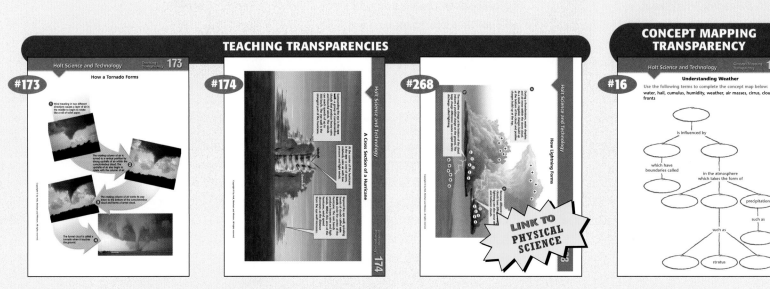

#173 — Holt Science and Technology — Teaching Transparency 173 — How a Tornado Forms

#174 — Holt Science and Technology — A Cross Section of a Hurricane — 174

#268 — Holt Science and Technology — How Lightning Forms — LINK TO PHYSICAL SCIENCE

CONCEPT MAPPING TRANSPARENCY

#16 — Holt Science and Technology — Concept Mapping Transparency 16 — Understanding Weather

Use the following terms to complete the concept map below: water, hail, cumulus, humidity, weather, air masses, cirrus, clouds, fronts

Meeting Individual Needs

DIRECTED READING

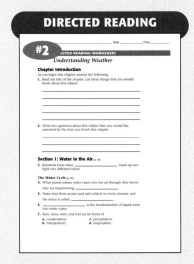

#2 — DIRECTED READING WORKSHEET

Understanding Weather

Chapter Introduction

As you begin this chapter, answer the following.

1. Read the title of the chapter. List three things that you already know about this subject.

2. Write two questions about this subject that you would like answered by the time you finish this chapter.

Section 1: Water in the Air (p. 36)

3. Rainbows form when _____ break up sunlight into different colors.

The Water Cycle (p. 36)

4. When plants release water vapor into the air through their leaves they are experiencing _____.

5. Water that flows across land and collects in rivers, streams, and the ocean is called _____.

6. _____ is the transformation of liquid water into water vapor.

7. Rain, snow, sleet, and hail are all forms of
   a. condensation.   c. precipitation.
   b. transpiration.   d. evaporation.

REINFORCEMENT & VOCABULARY REVIEW

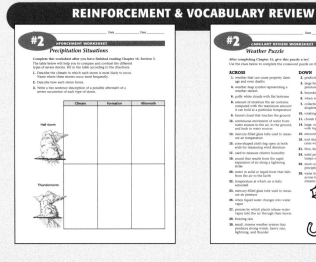

#2 — REINFORCEMENT WORKSHEET

Precipitation Situations

Complete this worksheet after you have finished reading Chapter 16, Section 3. The table below will help you to compare and contrast the different types of severe storms. Fill in the table according to the directions.

1. Describe the climate in which each storm is most likely to occur. Name where these storms occur most frequently.

2. Describe how each storm forms.

3. Write a two sentence description of a possible aftermath of a severe occurrence of each type of storm.

	Climate	Formation	Aftermath
Hail storm			
Thunderstorm			

#2 — VOCABULARY REVIEW WORKSHEET

Weather Puzzle

After completing Chapter 16, give this puzzle a try!
Use the clues below to complete the crossword puzzle on the next page.

ACROSS

1. weather that can cause property damage and even deaths
4. weather map symbol representing a weather station
6. puffy white clouds with flat bottoms
8. amount of moisture the air contains compared with the maximum amount it can hold at a particular temperature
9. funnel cloud that touches the ground
12. continuous movement of water from water sources to the air, to the ground, and back to water sources
13. mercury-filled glass tube used to measure air temperature
15. cone-shaped cloth bag open at both ends for measuring wind direction
17. used to measure relative humidity
19. sound that results from the rapid expansion of air along a lightning strike
20. water in solid or liquid form that falls from the air to the Earth
22. temperature at which air is fully saturated
23. mercury-filled glass tube used to measure air pressure
26. when liquid water changes into water vapor
27. process by which plants release water vapor into the air through their leaves
29. freezing rain
30. small, intense weather system that produces strong winds, heavy rain, lightning, and thunder

DOWN

2. prediction of weather conditions
3. large body of air that has similar temperature and moisture throughout
5. boundary between two air masses
7. when water vapor becomes a liquid
11. collection of millions of tiny water droplets or ice crystals
10. rotating cups that measure wind speed
11. clouds that form in layers
14. large, rotating tropical weather system with high-speed winds
16. amount of moisture in the air
18. tool shaped like an arrow that indicates wind direction
21. thin, feathery clouds at high altitudes
24. solid precipitation that falls as balls or lumps of ice
25. most common form of solid precipitation
28. water from precipitation that flows across land and collects in rivers and streams

SCIENCE PUZZLERS, TWISTERS & TEASERS

#2 — SCIENCE PUZZLERS, TWISTERS & TEASERS

Understanding Weather

Microclimates

1. Decide which two-letter air mass symbol best represents each of the "microclimates" below. The symbols are mP, cP, mT, & cT. Write the appropriate symbol in the space provided.

The air mass over . . .

a. a steaming cup of tomato soup _____

b. a glass of iced tea _____

c. a hot bubble bath _____

d. a lamp _____

e. an ice cube tray _____

Poetic Fronts

2. Use the Poetic Precipitation Reporter's poems to classify the following fronts. Assume the air is moving from the left to right.

a. Warm air on the left
   Cold air on the right
   There should have been a rainstorm
   But a cold front caused a theft

b. Warm air on the left
   Cold air on the right
   Brought a little rain
   And it was warm all night

c. To the right was warm air
   To the left was cold
   When I went outside
   An umbrella I did hold

Review & Assessment

STUDY GUIDE

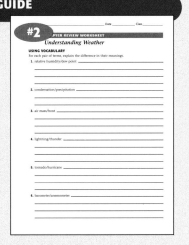

CHAPTER TESTS WITH PERFORMANCE-BASED ASSESSMENT

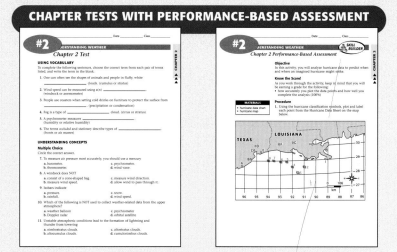

Lab Worksheets

INQUIRY LABS

ECOLABS & FIELD ACTIVITIES

LONG-TERM PROJECTS & RESEARCH IDEAS

WHIZ-BANG DEMONSTRATIONS

DATASHEETS FOR LABBOOK

#2 Watching the Weather

#2 Let It Snow!

#2 Gone with the Wind

Applications & Extensions

CRITICAL THINKING & PROBLEM SOLVING

EYE ON THE ENVIRONMENT

SECTION 1

Water in the Air

▶ **Earth's Water Cycle**

Although the atmosphere contains only about 0.001 percent of the total volume of water on the planet (about 1.46×10^9 km³), it is an essential link between land masses and bodies of water on Earth.

- The rate at which water evaporates into Earth's atmosphere is about 5.1×10^{17} L per year.

- About 78 percent of all precipitation falls over Earth's oceans. Of the 22 percent that falls on land, about 65 percent returns to the air by evaporation.

▶ **Clouds**

Clouds may be composed of water droplets, ice crystals, or a combination of the two. For example, cirrus clouds are made of only ice crystals; stratus clouds are made of only water droplets; and altostratus clouds are mixtures of ice and liquid water. Cumulonimbus clouds, which produce snowflakes and hail, consist of water droplets near the bottom of the clouds and ice crystals in the upper parts of the clouds.

▶ **Precipitation**

Because of differences in condensation rates within the cloud, the millions of droplets of water that make up a cloud are not all the same size. Larger drops collide and merge with smaller drops to form raindrops.

IS THAT A FACT!

➤ Hailstones as big as grapefruits have fallen to Earth during some severe storms.

SECTION 2

Air Masses and Fronts

▶ **Fronts**

A weather front separating two air masses always slopes upward over a colder air mass because the colder air is denser.

- As a warm front approaches, the first clouds to appear in the sky are the high clouds: cirrus, cirrostratus, and cirrocumulus. As the front moves closer, medium-height clouds appear. As the front moves even closer, low clouds appear. As the front arrives, the temperature and air pressure drop, and in the Northern Hemisphere, winds blow from the northeast. Nimbostratus clouds bring drizzly precipitation, which may fall within 24 hours of the first cloud sighting.

- When a cold front enters an area, cumulonimbus clouds produce thunderstorms, heavy rain, or snow along the front. After the cold front passes through an area, winds change direction and barometric pressure rises. Behind the front, temperatures usually fall, bringing cool, clear weather to the area.

SECTION 3

Severe Weather

▶ Tornadoes

Meteorologists rate tornado intensity using the Fujita Tornado Intensity Scale. An F0 tornado is a relatively weak storm that may damage chimneys, tree branches, and billboard signs. An F1 tornado is a moderate storm that can peel the surfaces off roofs, overturn mobile homes, and push moving cars off roads. F2 and F3 tornadoes cause considerable to severe damage by tearing roofs off houses, overturning railroad cars, and uprooting mature trees. An F4 tornado is a devastating storm that levels houses and other buildings and tosses cars into the air. The most severe tornado is an F5 storm, which can lift houses off their foundations and carry them great distances. An F5 storm can carry cars over 100 m and strip the bark off trees.

▶ Hurricanes

On the Saffir-Simpson scale, hurricanes fall into five categories. Category 1 hurricanes have sustained winds between 119 and 153 km/h and usually cause relatively minimal damage. Category 2 hurricanes cause moderate damage with winds ranging between 154 and 177 km/h. Category 3 hurricanes cause extensive damage with winds that blow between 178 and 209 km/h. Category 4 hurricanes, like Hurricane Andrew, which struck Florida in 1992, have sustained winds between 210 and 250 km/h. Category 5 hurricanes, classified as catastrophic storms, have sustained winds of more than 250 km/h.

IS THAT A FACT!

➤ A hurricane is called a *willy-willy* in Australia, a *taino* in Haiti, a *baguio* in the Philippines, and a *cordonazo* in western Mexico.

SECTION 4

Forecasting the Weather

▶ Weather-Prediction Methods

One of the simplest methods for weather prediction, the persistence method, assumes that the atmospheric conditions at the time of a weather forecast will not change in the near future. It is fairly accurate in areas where weather patterns change very slowly, such as in southern California, where summer weather typically changes very little from day to day. Other methods include the following:

- The trends method involves determining high and low pressure areas, gauging the velocity of weather fronts, and locating areas of clouds and precipitation. A forecaster then uses this data to predict where these weather phenomena will be in the future. This method of weather prediction works well only when weather systems maintain constant velocities for a long period of time.

- The climatology method involves averaging weather data that has accumulated over many years to make a forecast. This method is accurate when weather patterns are similar to those expected for a given time of year.

- The numerical weather-prediction (NWP) method uses complex computer programs to generate models of probable air temperature, barometric pressure, wind velocity, and precipitation. A meteorologist then analyzes how he or she thinks the features predicted by the computer will interact to produce the day's weather. One shortcoming of this method is that it requires very accurate, comprehensive data. If initial weather conditions are not known, the prediction of how the system will change might not be accurate. Despite its flaws, the NWP method is one of the most reliable methods available.

For background information about teaching strategies and issues, refer to the *Professional Reference for Teachers*.

CHAPTER 2

Understanding Weather

Pre-Reading Questions

Students may not know the answers to these questions before reading the chapter, so accept any reasonable response.

Suggested Answers

1. Answers will vary.

2. Weather is caused by the movement and interaction of air masses.

CHAPTER 2

Understanding Weather

Sections

Pre-Reading Questions

1. Name some different kinds of clouds. How are they different?
2. What causes weather?

34

TWISTING TERROR

North America experiences an average of 700 tornadoes per year—more tornadoes than any other continent. Most of these tornadoes hit an area in the central United States called Tornado Alley. Tornado Alley has more tornadoes than any other area because its flatness and location on the Earth's surface make it possible for warm air masses and cold air masses to collide. In this chapter, you will learn about what causes weather and how weather can suddenly turn violent.

---

 internet**connect**

 HRW On-line Resources

go.hrw.com
For worksheets and other teaching aids, visit the HRW Web site and type in the keyword: **HSTWEA**

SciLINKS
NSTA

www.scilinks.com
Use the sciLINKS numbers at the end of each chapter for additional resources on the **NSTA** Web site.

Smithsonian Institution

www.si.edu/hrw
Visit the Smithsonian Institution Web site for related on-line resources.

 **CNNfyi.com**

www.cnnfyi.com
Visit the CNN Web site for current events coverage and classroom resources.

START-UP Activity

A MEETING OF THE MASSES

In this activity, you will model what happens when two air masses with different temperature characteristics meet.

Procedure

1. Fill a **beaker** with **500 mL of cooking oil.** Fill another **beaker** with **500 mL of water.** The cooking oil represents a less dense warm air mass. The water represents a denser cold air mass.

2. Predict what would happen if you tried to mix the two liquids together. Record your prediction in your ScienceLog.

3. Pour the contents of each beaker into a **clear, plastic rectangular container** at the same time from opposite ends of the container.

4. Observe what happens when the oil and water meet. Record your observations in your ScienceLog.

Analysis

5. What happens when the different liquids meet?

6. Does the prediction you made in step 2 match your results?

7. Based on your results, hypothesize what would happen if a cold air mass met a warm air mass. Record your hypothesis in your ScienceLog.

35

A MEETING OF THE MASSES

MATERIALS
FOR EACH GROUP:
• two 750 mL beakers
• 500 mL cooking oil
• 500 mL water
• clear rectangular container

Answers to START-UP Activity

5. The oil floats on the water.

6. Answers will vary.

7. Answers will vary.

Focus

Water in the Air

This section begins with a discussion of the water cycle. It then covers humidity and the process of condensation. Students will learn how to determine relative humidity. Finally, the section discusses cloud types and the forms of precipitation.

🔔 Bellringer

Place two glasses on your desk for students to observe: one filled with ice water and one filled with warm water. Put three drops of food coloring in each glass. Ask students why water droplets form on the outside of the cold container. Does the water seep through the glass? Does it come from the air? Why don't the water beads form on the warm container?

1) Motivate

GROUP ACTIVITY

Divide the class in half: half the students will pretend to be air molecules, and half will be water molecules. Ask the air molecules to stand two shoulder lengths apart in a square grid. Then have the water molecules stand between the air molecules without touching anyone. Tell students that they are modeling a warm air mass. Cool the air mass by moving the air molecules so that they are shoulder length apart. Some of the water molecules will be expelled as "precipitation." Finally, ask the air molecules to stand with their shoulders touching, showing how a cold air mass expels nearly all of the water molecules.

Terms to Learn

weather	condensation
water cycle	dew point
humidity	cloud
relative humidity	precipitation

What You'll Do

- ◆ Explain how water moves through the water cycle.
- ◆ Define *relative humidity*.
- ◆ Explain what the dew point is and its relation to condensation.
- ◆ Describe the three major cloud forms.
- ◆ Describe the four major types of precipitation.

Teaching Transparency 168 "The Water Cycle"

Directed Reading Worksheet Section 1

Water in the Air

There might not be a pot of gold at the end of a rainbow, but rainbows hold another secret that you might not be aware of. Rainbows are evidence that the air contains water. Water droplets break up sunlight into the different colors that you can see in a rainbow. Water can exist in the air as a solid, liquid, or gas. Ice, a solid, is found in clouds as snowflakes. Liquid water exists in clouds as water droplets. And water in gaseous form exists in the air as water vapor. Water in the air affects the weather. **Weather** is the condition of the atmosphere at a particular time and place. In this section you will learn how water affects the weather.

The Water Cycle

Water in liquid, solid, and gaseous states is constantly being recycled through the water cycle. The **water cycle** is the continuous movement of water from water sources, such as lakes and oceans, into the air, onto and over land, into the ground, and back to the water sources. Look at **Figure 1** below to see how water moves through the water cycle.

Figure 1 *In the water cycle, water is returned to the Earth's surface through precipitation.*

Condensation occurs when water vapor cools and changes back into liquid droplets. This is how clouds form.

Evaporation occurs when liquid water changes into water vapor, which is a gas.

Transpiration is the process by which plants release water vapor into the air through their leaves.

Precipitation occurs when rain, snow, sleet, or hail falls from the clouds onto the Earth's surface.

Runoff is water, usually from precipitation, that flows across land and collects in rivers, streams, and eventually the ocean.

36

CONNECT TO
PHYSICAL SCIENCE

At the Earth's surface and in the atmosphere, energy from the sun moves water through the water cycle. Sunlight provides the energy for evaporation, transpiration, condensation, and precipitation. Have groups work together to create a poster that shows how energy from the sun drives the water cycle.

Humidity

Have you ever spent a long time styling your hair before school and had a bad hair day anyway? You walked outside and—wham—your straight hair became limp, or your curly hair became frizzy. Most bad hair days can be blamed on humidity. **Humidity** is the amount of water vapor or moisture in the air. And it is the moisture in the air that makes your hair go crazy, as shown in **Figure 2.**

As water evaporates, the humidity of the air increases. But air's ability to hold water vapor depends on air temperature. As temperature increases, the air's ability to hold water also increases. **Figure 3** shows the relationship between air temperature and air's ability to hold water.

Figure 2 When there is more water in the air, your hair absorbs moisture and becomes longer.

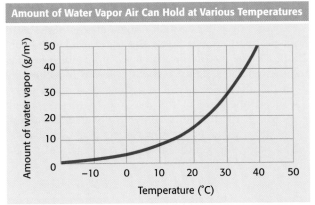

Amount of Water Vapor Air Can Hold at Various Temperatures

Figure 3 This graph shows that warmer air can hold more water vapor than cooler air.

Relative Humidity **Relative humidity** is the amount of moisture the air contains compared with the maximum amount it can hold at a particular temperature. Relative humidity is given as a percentage. When air holds all the water it can at a given temperature, the air is said to be *saturated.* Saturated air has a relative humidity of 100 percent. But how do you find the relative humidity of air that is not saturated? If you know the maximum amount of water vapor air can hold at a particular temperature and you know how much water vapor the air is actually holding, you can calculate the relative humidity.

Suppose that 1 m³ of air at a certain temperature can hold 24 g of water vapor. However, you know that the air actually contains 18 g of water vapor. You can calculate the relative humidity using the following formula:

$$\frac{\text{(present) } 18 \text{ g/m}^3}{\text{(saturated) } 24 \text{ g/m}^3} \times 100 = \text{(relative humidity) } 75\%$$

$\div\ 5\ \div\ \ ^\Omega\ _\leq\ ^\infty\ _{+\Omega}\ ^\surd\ 9\ _\infty^\leq\ \Sigma\ 2$
$+$

MATH BREAK

Relating Relative Humidity
Assume that a sample of air 1 m³ at 25°C, contains 11 g of water vapor. Calculate the relative humidity of the air using the value for saturated air shown in Figure 3.

37

CONNECT TO
LIFE SCIENCE

When the air is humid, hair becomes frizzy. Hair is made of a protein called keratin. Each hair fiber has a scaly outer cuticle, which you can feel by running your fingers up and down a single hair. The scales allow moisture to enter the inner part of the hair fiber. When the air is humid, hair absorbs moisture and becomes longer, making it frizzy. Hair dries out and becomes shorter when the air is dry. Because humidity can cause hair length to change by as much as 2.5 percent, a device called a hair hygrometer can very accurately measure changes in humidity. Have students design and build their own hair hygrometers.

Answer to Self-Check

Evaporation occurs when liquid water changes into water vapor and returns to the air. Humidity is the amount of water vapor in the air.

INDEPENDENT PRACTICE

After students have read this page, have them complete the following sentences:

- If the humidity is low, a _____ amount of water will evaporate from a wet-bulb thermometer and the _____ between the wet-bulb reading and the dry-bulb reading of the psychrometer will be high. (large, temperature difference)

- If the dry bulb reads 10°C, and the difference between the thermometers is 8°C, the relative humidity is _____. (15 percent)

MEETING INDIVIDUAL NEEDS

Advanced Learners Have small groups develop an activity to determine the relative humidity at different locations on the same day. Tell them to use the following materials: two identical thermometers, gauze, string, and room-temperature water. Suggest that they create a wet-bulb thermometer by tying a small piece of gauze to the bottom of one thermometer and saturating the covered end with water. Have students record their data in a table. Their data should include information about location, observable weather conditions, time of day, and the duration of the trial.

✓ Self-Check

How does humidity relate to the water cycle? *(Turn to page 136 to check your answer.)*

Water Vapor Versus Temperature If the temperature stays the same, relative humidity changes as water vapor enters or leaves the air. The more water vapor that is in the air at a particular temperature, the higher the relative humidity is. Relative humidity is also affected by changes in temperature. If the amount of water vapor in the air stays the same, the relative humidity decreases as the temperature rises and increases as the temperature drops.

Measuring Relative Humidity A *psychrometer* (sie KRAHM uht uhr) is an instrument used to measure relative humidity. It consists of two thermometers. One thermometer is called a wet-bulb thermometer. The bulb of this thermometer is covered with a damp cloth. The other thermometer is a dry-bulb thermometer. The dry-bulb thermometer measures air temperature.

As air passes over the wet-bulb thermometer, the water in the cloth begins to evaporate. As the water evaporates from the cloth, energy is transferred away from the wet-bulb and the thermometer begins to cool. If there is less humidity in the air, the water will evaporate more quickly and the temperature of the wet-bulb thermometer will drop. If the humidity is high, only a small amount of water will evaporate from the wet-bulb thermometer and there will be little change in temperature.

Follow the Numbers

Relative Humidity (in percentage)								
Dry-bulb reading (°C)	Difference between wet-bulb reading and dry-bulb reading (°C)							
	1	2	3	4	5	6	7	8
0	81	64	46	29	13			
2	84	68	52	37	22	7		
4	85	71	57	43	29	16		
6	86	73	60	48	35	24	11	
8	87	75	63	51	40	29	19	8
10	88	77	66	55	44	34	24	15
12	89	78	68	58	48	39	29	21
14	90	79	70	60	51	42	34	26
16	90	81	71	63	54	46	38	30
18	91	82	73	65	57	49	41	34
20	91	83	74	66	59	51	44	37

Relative humidity can be determined using a table such as this one. Locate the column that shows the difference between the wet-bulb and dry-bulb readings. Then locate the row that lists the temperature reading on the dry-bulb thermometer. The value where the column and row intersect is the relative humidity.

🌐 Multicultural CONNECTION

Before modern weather instruments were invented, natives on the Chiloé Islands, off the coast of Chile, used shells of the crab *Lithodes antarcticus* to measure relative humidity. A dry shell, normally light gray in color, shows red patches when humidity increases. It will become completely red if the humidity continues to rise, as during the rainy season. Australian Aborigines used dry kelp to predict rain. Some kelps contain magnesium chloride, which absorbs water vapor from the air. The kelp will feel damp long before it actually begins to rain.

The difference in temperature readings between the wet-bulb and dry-bulb thermometers indicates the amount of water vapor in the air. A larger difference between the two readings indicates that there is less water vapor in the air and thus lower humidity.

The Process of Condensation

You have probably seen water droplets form on the outside of a glass of ice water, as shown in **Figure 4.** Did you ever wonder where those water droplets came from? The water came from the surrounding air, and droplets formed because of condensation. **Condensation** is the process by which a gas, such as water vapor, becomes a liquid. Before condensation can occur, the air must be saturated; it must have a relative humidity of 100 percent. Condensation occurs when saturated air cools further.

Figure 4 *Condensation occurred when the air next to the glass cooled to below its dew point.*

Dew Point Air can become saturated when water vapor is added to the air through evaporation or transpiration. Air can also become saturated, as in the case of the glass of ice water, when it cools to its dew point. The **dew point** is the temperature to which air must cool to be completely saturated. The ice in the glass of water causes the air surrounding the glass to cool to its dew point.

Before it can condense, water vapor must also have a surface to condense on. On the glass of ice water, water vapor condenses on the sides of the glass. Another example you may already be familiar with is water vapor condensing on grass, forming small water droplets called *dew*.

SECTION REVIEW

1. What is the difference between humidity and relative humidity?

2. What are two ways that air can become saturated with water vapor?

3. What does a relative humidity of 75 percent mean?

4. How does the water cycle contribute to condensation?

5. **Analyzing Relationships** What happens to relative humidity as the air temperature drops below the dew point?

Out of Thin Air

1. Take a **plastic container,** such as a jar or drinking glass, and fill it almost to the top with room-temperature **water.**

2. Observe the outside of the can or container. Record your observations.

3. Add one or two **ice cubes,** and watch the outside of the container for any changes.

4. What happened to the outside of the container?

5. What is the liquid?

6. Where did the liquid come from? Why?

TRY at HOME

39

READING 📖 STRATEGY

Activity After students read this page, have them arrange the following steps in logical order:

• Water vapor condenses on smoke, dust, salt, and other small particles suspended in air. (4)

• Water vapor is added to the air. (2)

• Warm air rises and cools. (1)

• Air eventually becomes saturated. (3)

• Millions of droplets of liquid water collect to form a cloud. (5)

USING THE FIGURE

Have students carefully study **Figures 5, 6,** and **7.** Challenge them to determine one method of cloud classification based on their observations of the photographs. (In addition to being classified by altitude, as described on the following page, clouds are also classified according to shape.)

MATH and **MORE**

How heavy is a cloud? The first step in answering this question is to determine the cloud's volume. If an average cumulus cloud is 1,000 m long, 1,000 m wide, and 1,000 m tall, it has a volume of one billion cubic meters. Have students multiply that volume by the weight of water in 1 m^3 of a cumulus cloud (0.5 g). (1,000,000,000 m^3 × 0.5 g = 500,000,000 g, or 500,000 kg)

Clouds

Some look like cotton balls, some look like locks of hair, and others look like blankets of gray blocking out the sun. But what *are* clouds and how do they form? And why are there so many different-looking clouds? A **cloud** is a collection of millions of tiny water droplets or ice crystals. Clouds form as warm air rises and cools. As the rising air cools, it becomes saturated. At saturation the water vapor changes to a liquid or a solid depending on the air temperature. At higher temperatures, water vapor condenses on small particles, such as dust, smoke, and salt, suspended in the air as tiny water droplets. At temperatures below freezing, water vapor changes directly to a solid, forming ice crystals.

Cumulus

Figure 5 *Cumulus clouds look like piles of cotton balls.*

Stratus

Figure 6 *Although stratus clouds are not as tall as cumulus clouds, they cover more area.*

Cumulus Clouds Puffy, white clouds that tend to have flat bottoms, as shown in **Figure 5,** are called *cumulus clouds*. Cumulus clouds form when warm air rises. These clouds generally indicate fair weather. However, when these clouds get larger they produce thunderstorms. A cumulus cloud that produces thunderstorms is called a *cumulonimbus cloud*. When *-nimbus* or *nimbo-* is part of a cloud's name, it means that precipitation might fall from the cloud.

Stratus Clouds Clouds that form in layers, as shown in **Figure 6,** are called *stratus clouds*. Stratus clouds cover large areas of the sky, often blocking out the sun. These clouds are caused by a gentle lifting of a large body of air into the atmosphere. *Nimbostratus clouds* are dark stratus clouds that usually produce light to heavy, continuous rain. When water vapor condenses near the ground, it forms a stratus cloud called *fog*.

40

Homework

PORTFOLIO

Cloud Models On a poster board, have students use cotton balls to make models of different types of clouds at different altitudes. Students should create labels to describe the clouds and the types of weather they are associated with. Sheltered English

MISCONCEPTION //// ALERT

What appears to be white smoke from an airplane's engine is not smoke at all. Condensation trails, or *contrails*, form as the combustion of the aircraft's fuel causes water vapor to condense and freeze along the airplane's exhaust tail. A thick contrail that will not dissipate is a sign that a frontal system is approaching.

Cirrus Clouds As you can see in **Figure 7,** *cirrus* (SIR uhs) *clouds* are thin, feathery, white clouds found at high altitudes. Cirrus clouds form when the wind is strong. Cirrus clouds may indicate approaching bad weather if they thicken and lower in altitude.

Clouds are also classified by the altitude at which they form. The illustration in **Figure 8** shows the three altitude groups used to categorize clouds.

Figure 7 *Cirrus clouds are made of ice crystals.*

Figure 8 Cloud Types Based on Form and Altitude

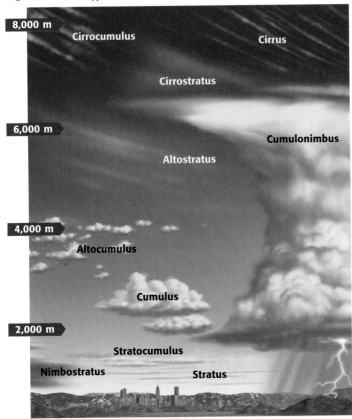

8,000 m

Cirrocumulus

Cirrus

Cirrostratus

Cumulonimbus

6,000 m

Altostratus

4,000 m

Altocumulus

Cumulus

2,000 m

Stratocumulus

Nimbostratus

Stratus

High Clouds
Because of the cold temperatures at high altitude, high clouds are made up of ice crystals. The prefix *cirro-* is used to describe high clouds.

Middle Clouds
Middle clouds can be made up of both water droplets and ice crystals. The prefix *alto-* is used to describe middle clouds.

Low Clouds
Low clouds are made up of water droplets. The prefix *strato-* is commonly used to describe these types of clouds.

41

CROSS-DISCIPLINARY FOCUS

Language Arts Recording descriptive and useful observations of clouds is an essential skill in amateur meteorology. Show students examples of good and bad cloud descriptions, and discuss the factors that make scientific observations useful. When describing clouds, students should consider these questions: Do they appear close to the ground or high up? Are they white, light gray, or dark gray? Are they flat and sheetlike, rounded and fluffy, or thin and wispy? Are the clouds distinct? Have students make several entries in their weather journal that describe the clouds they observe over the period of a week. Entries should include both descriptions and sketches, and students should write a weather forecast based on the clouds observed each day.

Explain that a water molecule has a positive end and a negative end. Opposite charges attract, so the positive end of one water molecule attracts the negative end of another. This attraction helps explain why small water droplets that collide are able to form relatively large raindrops. Large raindrops are deformed by air pressure as they fall through the atmosphere. This causes them to be shaped like a hamburger bun with a concave bottom, as shown in **Figure 9.**

DEMONSTRATION

Hail Formation Melt three or four different-colored crayons or candles in separate containers over a hot plate. Dip a thick, weighted string into one color of wax, blow it dry, and repeat with each different color. After you have built up several layers, cut the wax widthwise. Display the concentric ringed formation to the class. Ask students what kind of precipitation forms in a similar manner. (hail)
Sheltered English

GOING FURTHER

 Meteorologists sometimes use a technique known as cloud seeding to cause or increase precipitation. Have students research and write a report about this technique. After they have gathered their information, have small groups debate the pros and cons of artificially stimulating precipitation.

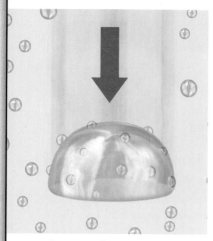

Figure 9 *Cloud droplets get larger by colliding and joining with other droplets. When the water droplets become too heavy, they fall as precipitation.*

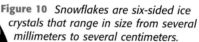
Figure 10 *Snowflakes are six-sided ice crystals that range in size from several millimeters to several centimeters.*

Precipitation

Water vapor that condenses to form clouds can eventually fall to the ground as precipitation. **Precipitation** is water, in solid or liquid form, that falls from the air to the Earth. There are four major forms of precipitation—rain, snow, sleet, and hail.

Rain, the most common form of precipitation, is liquid water that falls from the clouds to Earth. A cloud produces rain when its water droplets become large enough to fall. A cloud droplet begins as a water droplet smaller than the period at the end of this sentence. Before a cloud droplet falls as precipitation, it must increase in size to about 100 times its normal diameter. **Figure 9** illustrates how a water droplet increases in size until it is finally large enough to fall as precipitation.

Snow and Sleet The most common form of solid precipitation is *snow*. Snow forms when temperatures are so cold that water vapor changes directly to a solid. Snow can fall as individual ice crystals or combine to form snowflakes, like the one shown in **Figure 10.**

Sleet, also called freezing rain, forms when rain falls through a layer of freezing air. The rain freezes, producing falling ice. Sometimes rain does not freeze until it hits a surface near the ground. When this happens, the rain changes into a layer of ice called *glaze,* as shown in **Figure 11.**

Figure 11 *Glaze ice forms as rain freezes on surfaces near the ground.*

42

Multicultural
CONNECTION

Have students research the rainmakers in Hopi Indian culture. Students might find out about *Leenangkatsina,* whose flute brings rain; *Qaleetaqa,* who carries lightning and a bull-roarer to bring rain; and *Si'o Sa'lakwmana* or *Pawtiwa,* both of whom bring rain and mist to villages.

WEIRD SCIENCE

The largest hailstone ever recorded fell on Coffeyville, Kansas, on September 3, 1970. The hailstone was the size of a softball, and weighed 3.7 kg.

Hail Solid precipitation that falls as balls or lumps of ice is called *hail*. Hail usually forms in cumulonimbus clouds. Updrafts of air in the clouds carry raindrops to high altitudes in the cloud, where they freeze. As the frozen raindrops fall, they collide and combine with water droplets. Another updraft of air can send the hail up again high into the cloud. Here the water drops collected by the hail freeze, forming another layer of frozen ice. If the upward movement of air is strong enough, the hail can accumulate many layers of ice. Eventually, the hail becomes too heavy and falls to the Earth's surface, as shown in **Figure 12.** Hail is usually associated with warm weather and most often occurs during the spring and summer months.

Figure 12 *The impact of large hailstones can damage property and crops.*

Measuring Precipitation A *rain gauge* is an instrument used to measure the amount of rainfall. A rain gauge typically consists of a funnel and a cylinder, as shown in **Figure 13.** Rain falls into the funnel and collects in the cylinder. Markings on the cylinder indicate how much rain has fallen.

Snow is measured by both depth and water content. The depth of snow is measured using a measuring stick. The snow's water content is determined by melting the snow and measuring the amount of water.

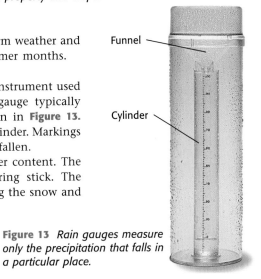

Funnel

Cylinder

Figure 13 *Rain gauges measure only the precipitation that falls in a particular place.*

SECTION REVIEW

1. How do clouds form?

2. Why are some clouds formed from water droplets, while others are made up of ice crystals?

3. Describe how rain forms.

4. **Applying Concepts** How can rain and hail fall from the same cumulonimbus cloud?

internet**connect**

SCI*LINKS*
NSTA

TOPIC: Collecting Weather Data
GO TO: www.scilinks.org
*sci*LINKS NUMBER: HSTE380

43

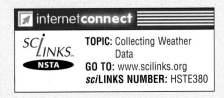

Focus

Air Masses and Fronts

In this section, students learn what air masses are and how they affect weather in the United States. Students also learn about the boundaries of air masses—known as *fronts*—and the types of weather associated with these boundaries.

Bellringer

Ask students to write down as many different qualities of air as possible. (Students might note that air can be humid or dry, hot or cold, or have a high pressure or a low pressure.) Tell students that the air they are breathing now was hundreds of miles away yesterday. Ask them to think about what caused that air to move. Explain that air masses tend to flow from areas of high pressure to areas of low pressure, just as the air inside a balloon escapes when the balloon is punctured.

1 Motivate

DEMONSTRATION

Students may have a difficult time understanding how hot and cold air masses stay separated as they move. Tell students that air at different temperatures has different densities. In this respect, air behaves much like water. To demonstrate how temperature can separate liquid masses, fill a large beaker or jar with hot water and a small beaker with cold water. Add several drops of blue food coloring to the cold water to make it visible. Slowly pour the cold water down the side of the jar. Encourage students to describe and explain what they observe.

Terms to Learn

air mass
front

What You'll Do

◆ Explain how air masses are characterized.
◆ Describe the four major types of air masses that influence weather in the United States.
◆ Describe the four major types of fronts.
◆ Relate fronts to weather changes.

Air Masses and Fronts

Have you ever wondered how the weather can change so fast? One day the sun is shining and you are wearing shorts, and the next day it is so cold you need a coat. Changes in weather are caused by the movement and interaction of air masses. An **air mass** is a large body of air that has similar temperature and moisture throughout. In this section you will learn about air masses and how their interaction influences the weather.

Air Masses

An air mass gets its moisture and temperature characteristics from the area over which it forms. These areas are called *source regions*. For example, an air mass that develops over the Gulf of Mexico is warm and wet because this area is warm and has a lot of water that evaporates into the air. There are many types of air masses, each associated with a particular source region. The characteristics of these air masses are represented on maps with a two-letter symbol, as shown in **Figure 14**. The first letter indicates the moisture characteristics of the air mass, and the second symbol represents the temperature characteristics of the air mass.

Figure 14 *This map shows the source regions for air masses that influence weather in North America.*

maritime (m)–forms over water; wet

continental (c)–forms over land; dry

polar (P)–forms over the polar regions; cold

tropical (T)–develops over the Tropics; warm

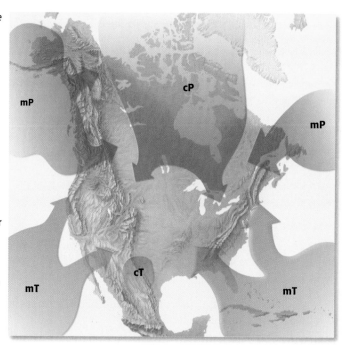

MISCONCEPTION ALERT

People often assume that humid air is heavier than dry air. Actually, humid air rises like a balloon because it is less dense than dry air at the same pressure and temperature. The reason is that water molecules are lighter than N_2 and O_2, the main constituents of air. The more water vapor there is in a mass of air, the more N_2 and O_2 is displaced. In general, when humidity increases, air becomes less dense and rises. As an air mass rises, it cools, the water vapor condenses, and the air mass eventually sinks. If water vapor were heavier than air, clouds would only form at the Earth's surface.

Cold Air Masses Most of the cold winter weather in the United States is influenced by three polar air masses. A continental polar air mass develops over land in northern Canada. In the winter, this air brings extremely cold weather to the United States, as shown in **Figure 15.** In the summer, it generally brings cool, dry weather.

A maritime polar air mass that forms over the North Pacific Ocean mostly affects the Pacific Coast. This air mass is very wet, but not as cold as the air mass that develops over Canada. In the winter, this air mass brings rain and snow to the Pacific Coast. In the summer, it brings cool, foggy weather.

A maritime polar air mass that forms over the North Atlantic Ocean usually affects New England and eastern Canada. In the winter, it produces cold, cloudy weather with precipitation. In the summer, the air mass brings cool weather with fog.

Figure 15 *A cP air mass generally moves southeastward across Canada and into the northern United States.*

Warm Air Masses Four warm air masses influence the weather in the United States. A maritime tropical air mass that develops over warm areas in the North Pacific Ocean is lower in moisture content and weaker than the maritime polar air mass. As a result, southern California receives less precipitation than the rest of California.

Other maritime tropical air masses develop over the warm waters of the Gulf of Mexico and the North Atlantic Ocean. These air masses move north across the East Coast and into the Midwest. In the summer, they bring hot and humid weather, thunderstorms, and hurricanes, as shown in **Figure 16.** In the winter, they bring mild, often cloudy weather.

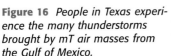
Figure 16 *People in Texas experience the many thunderstorms brought by mT air masses from the Gulf of Mexico.*

45

IS THAT A FACT!

Air masses can extend upward for thousands of meters and can reach the top of the troposphere—an altitude of 10 to 16 km!

CONNECT TO
LIFE SCIENCE

Why do people complain of aching joints before a thunderstorm? A study found that nearly 75 percent of arthritis sufferers felt more pain in their joints when air pressure was falling. Although this effect has been thoroughly documented, there is no definitive evidence of why it occurs.

2 Teach

DISCUSSION

Air Masses and You Have students use **Figure 14** to determine which type of air mass is mainly responsible for the weather in your area. Have students describe the general temperatures and humidity typical of your area. Then have students compare their observations with the information in this section.

Multicultural
CONNECTION

Local weather patterns are heavily influenced by air masses, which tend to bring predictable weather. All cultures have names for familiar weather patterns. For example, in Tunisia, Africa, weather forecasters often predict "hot and *chili*" conditions. This forecast may not make sense to people elsewhere, but to a Tunisian, *chili* refers to a hot wind blowing from the North African desert. Similarly, in parts of the eastern United States, people refer to the hot, dry, and relatively windless weeks of August as the Indian summer. Have interested students research the names and characteristics of typical weather patterns in other countries.

Directed Reading Worksheet Section 2

Teaching Transparency 170 "Air Masses in North America"

READING STRATEGY

Activity As they learn about the different types of fronts, have students draw each front in their ScienceLog. Students should label each diagram with arrows indicating the direction the air masses are moving and write captions describing the type of weather that is associated with the front.

MEETING INDIVIDUAL NEEDS

Learners Having Difficulty

Perform the following demonstration to show students how cold and warm fronts form. Obtain a pair of surgical gloves. Use magic markers to color one glove red and the other glove blue. When the gloves are completely dry, put them on. Tell students that the blue glove represents a cold air mass and the red glove represents a warm air mass. To show how a cold front forms, hold your hands in front of you, and move your "blue hand" toward your "red hand." Just before they touch, slide your blue hand under your red hand and push your red hand up. To simulate the formation of a warm front, keep your blue hand stationary, and move your red hand toward your blue hand. As your hands touch, slide your red hand up and over your blue hand. Sheltered English

Teaching Transparency 171
"Cold and Warm Fronts"

BRAIN FOOD

The term *front* was first used to describe weather systems during World War I in Europe. Meteorologists in Norway thought the boundaries between different air masses were much like the opposing armies on the battle front.

Cold Front

A cold air mass meets and displaces a warm air mass. Because the moving cold air is more dense, it moves under the less-dense warm air, pushing it up.

Cold fronts can move fast, producing thunderstorms, heavy rain, or snow. Cooler weather usually follows a cold front because the warm air is pushed away from the Earth's surface.

Warm Front

A warm air mass meets and overrides a cold air mass. The warm, less-dense air moves over the cold, denser air. The warm air gradually replaces the cold air.

Warm fronts generally bring drizzly precipitation. After the front passes, weather conditions are clear and warm.

A continental tropical air mass forms over the deserts of northern Mexico and the southwestern United States. This air mass influences weather in the United States only during the summer. It generally moves northeastward, bringing clear, dry, and very hot weather.

Fronts

Air masses with different characteristics, such as temperature and humidity, do not usually mix. So when two different air masses meet, a boundary forms between them. This boundary is called a **front**. Weather at a front is usually cloudy and stormy. The four different types of fronts—cold fronts, warm fronts, occluded fronts, and stationary fronts—are illustrated on these two pages. Fronts are usually associated with weather in the middle latitudes, where there are both cold and warm air masses. Fronts do not occur in the Tropics because only warm air masses exist there.

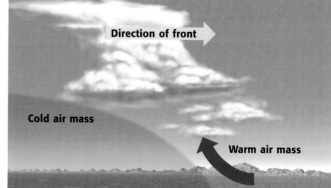

Direction of front

Cold air mass

Warm air mass

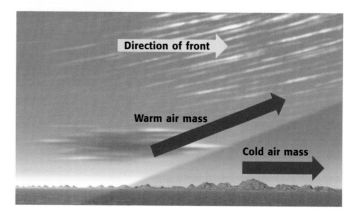

Direction of front

Warm air mass

Cold air mass

CROSS-DISCIPLINARY FOCUS

History During World War I, European nations stopped broadcasting weather reports, fearing that they would be used by advancing enemy troops. This left nonaligned countries such as Norway to develop their own meteorology program. Norwegian meteorologists responded by forming the famous Bergen School, which greatly advanced the field of meteorology. They discovered distinct air masses in the atmosphere and found that these masses traveled with the winds. Influenced by the war, the meteorologists described air masses using military terms. They imagined Europe as a battleground where different air masses fought like armies, each trying to advance on the other. The boundary between the air masses, where the "battle" occurs, was called the front.

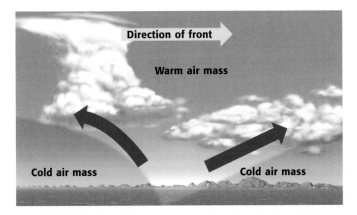

Direction of front

Warm air mass

Cold air mass

Cold air mass

Occluded Front

A faster-moving cold air mass overtakes a slower-moving warm air mass and forces the warm air up. The cold air mass then continues advancing until it meets a cold air mass that is warmer. The cold air mass then forces this air mass to rise.

An occluded front has cool temperatures and large amounts of precipitation.

Cold air mass

Warm air mass

Stationary Front

A cold air mass meets a warm air mass and little horizontal movement occurs.

The weather associated with a stationary front is similar to that produced by a warm front.

SECTION REVIEW

1. What are the characteristics that define air masses?

2. What are the major air masses that influence the weather in the United States?

3. What are fronts, and what causes them?

4. What kind of front forms when a cold air mass displaces a warm air mass?

5. **Analyzing Relationships**   Explain why the Pacific Coast has cool, wet winters and warm, dry summers.

---

internetconnect

SCiLINKS.
NSTA

TOPIC: Air Masses and Fronts
GO TO: www.scilinks.org
*sci*LINKS NUMBER: HSTE385

47

▼ *Answers to Section Review*

1. temperature and moisture

2. maritime tropical, maritime polar, continental tropical, and continental polar

3. Fronts are boundaries that form between two different air masses. Boundaries form because air masses with different moisture and temperature characteristics do not mix easily.

4. a cold front

5. In the winter, the Pacific Coast's climate is governed by a maritime polar air mass that brings wet weather and cool temperatures. In the summer, the Pacific Coast's climate is governed by a maritime tropical air mass that brings warm temperatures and little moisture.

Focus

Severe Weather

In this section, students learn about the conditions that form thunderstorms, tornadoes, and hurricanes. The discussion of thunderstorms includes an explanation of lightning and thunder, while the discussion of tornadoes and hurricanes includes information on the incredible damage these storms cause.

Bellringer

Show students a picture of a thunderstorm. Then have them write a one-paragraph description of a thunderstorm in their ScienceLog. Tell them to focus on the characteristics that distinguish thunderstorms from other forms of weather. Ask them to describe the weather conditions immediately before, during, and after a thunderstorm.

1 Motivate

DEMONSTRATION

Perform this demonstration to simulate the forces that cause thunder. Inflate a balloon with air, and tie it closed. Explain that thunder occurs when lightning superheats air, causing the gases to expand rapidly. The air in the balloon is under pressure, so it will also expand rapidly if the pressure is suddenly released. The rapid expansion of air causes vibrations that we hear as sound. Hold up a pin or needle, pause, and pop the balloon with a flourish. **Sheltered English**

Directed Reading Worksheet Section 3

Terms to Learn

thunderstorm   tornado
lightning   hurricane
thunder

What You'll Do

◆ Explain what lightning is.
◆ Describe the formation of thunderstorms, tornadoes, and hurricanes.
◆ Describe the characteristics of thunderstorms, tornadoes, and hurricanes.

Severe Weather

Weather in the mid-latitudes can change from day to day. These changes result from the continual shifting of air masses. Sometimes a series of storms will develop along a front and bring severe weather. *Severe weather* is weather that can cause property damage and even death. Examples of severe weather include thunderstorms, tornadoes, and hurricanes. In this section you will learn about the different types of severe weather and how each type forms.

Thunderstorms

Thunderstorms, as shown in **Figure 17,** are small, intense weather systems that produce strong winds, heavy rain, lightning, and thunder. As you learned in the previous section, thunderstorms can occur along cold fronts. But that's not the only place they develop. There are only two atmospheric conditions required to produce thunderstorms: the air near the Earth's surface must be warm and moist, and the atmosphere must be unstable. The atmosphere is unstable when the surrounding air is colder than the rising air mass. As long as the air surrounding the rising air mass is colder, the air mass will continue to rise.

Thunderstorms occur when warm, moist air rises rapidly in an unstable atmosphere. When the warm air reaches its dew point, the water vapor in the air condenses, forming cumulus clouds. If the atmosphere is extremely unstable, the warm air will continue to rise, causing the cloud to grow into a dark, cumulonimbus cloud. These clouds can reach heights of more than 15 km.

Figure 17 *A typical thunderstorm produces approximately 470 million liters of water and enough electricity to provide power to the entire United States for 20 minutes.*

Trees sometimes explode when struck by lightning. Why? Lightning causes the sap in the tree to vaporize (turn from a liquid to a gas). The steam expands rapidly as it is heated, causing the tree to explode.

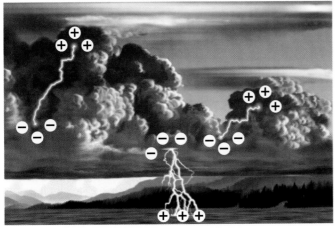
Figure 18 *The upper part of a cloud usually carries a positive electrical charge, while the lower part of the cloud carries mainly negative charges.*

Physics
CONNECTION

Have you ever wondered why you don't see lightning and hear thunder at the same time? Well, there's an easy explanation. Light travels faster than sound. The light reaches you almost instantly, but the sound travels only 1 km every 3 seconds. The closer the lightning is to where you are, the sooner you will hear the thunder.

Lightning Thunderstorms are very active electrically. **Lightning** is a large electrical discharge that occurs between two oppositely charged surfaces, as shown in **Figure 18.** Have you ever touched someone after scuffing your feet on the carpet and received a mild shock? If so, you have experienced how lightning forms. While walking around, friction between the floor and your shoes builds up an electrical charge in your body. When you touch someone else, the charge is released.

When lightning strikes, energy is released. This energy is transferred to the air and causes the air to expand rapidly and send out sound waves. **Thunder** is the sound that results from the rapid expansion of air along the lightning strike.

Severe Thunderstorms Severe thunderstorms produce one or more of the following conditions—high winds, hail, flash floods, and tornadoes. Hailstorms damage crops, dent the metal on cars, and break windows. Sudden flash flooding due to heavy rains causes millions of dollars in property damage annually and is the biggest cause of weather-related deaths.

Lightning, which occurs with all thunderstorms, is responsible for thousands of forest fires each year in the United States. Lightning also kills or injures hundreds of people a year in the United States.

Figure 19 *Lightning often strikes the highest object in an area.*

49

GROUP ACTIVITY

Making Models Have pairs of students work together to model a tornado vortex. Supply each pair with a clean, empty plastic jar with its lid, water, food coloring, a teaspoon of liquid dish soap, and a teaspoon of vinegar. Have students fill the jars about three-quarters full of water. Instruct them to add a few drops of food coloring, the soap, and the vinegar. Instruct them to cap the jar tightly and shake it vigorously. Once the solution is mixed, tell students to give the jars a quick twist with a flick of the wrist. Students will observe that a vortex will form and lengthen. Sheltered English

  BRAIN FOOD

Most tornadoes develop from thunderstorms at the leading edge of a cold front. Ask students to think about why this is so. (The cool air wedges under the warm air, forcing it to rise rapidly and become unstable. The distinct temperature and pressure differences between the two air masses greatly increase the chance of a tornado forming.)

USING SCIENCE FICTION

Encourage students to read "All Summer in a Day" by Ray Bradbury in the *Holt Anthology of Science Fiction*.

 Teaching Transparency 173 "How a Tornado Forms"

Tornadoes

Tornadoes are produced in only 1 percent of all thunderstorms. A **tornado** is a small, rotating column of air that has high wind speeds and low central pressure and that touches the ground. A tornado starts out as a funnel cloud that pokes through the bottom of a cumulonimbus cloud and hangs in the air. It is called a tornado when it makes contact with the Earth's surface. **Figure 20** shows the development of a tornado.

Figure 20 How a Tornado Forms

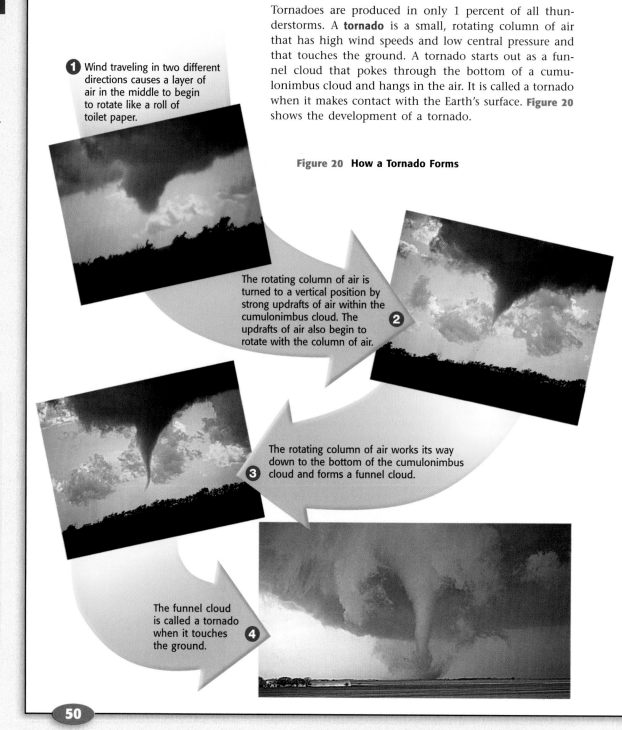

1 Wind traveling in two different directions causes a layer of air in the middle to begin to rotate like a roll of toilet paper.

2 The rotating column of air is turned to a vertical position by strong updrafts of air within the cumulonimbus cloud. The updrafts of air also begin to rotate with the column of air.

3 The rotating column of air works its way down to the bottom of the cumulonimbus cloud and forms a funnel cloud.

4 The funnel cloud is called a tornado when it touches the ground.

50

IS THAT A FACT!

Because of the Coriolis effect, almost all tornadoes in the Northern Hemisphere spin in a counterclockwise direction. In rare cases, a tornado will spin in a clockwise direction. These rare tornadoes are always paired with a tornado that rotates counterclockwise.

WEIRD SCIENCE

People have reported seeing "naked" chickens after tornadoes strike rural areas. A likely explanation is that tornadoes cause chickens to shed their feathers, or molt. Chickens often molt when attacked. As the chickens molt, the strong tornado winds blow their feathers off.

Twists of Terror About 75 percent of the world's tornadoes occur in the United States. The majority of these tornadoes happen in the spring and early summer when cold, dry air from Canada collides with warm, moist air from the Tropics. The length of a tornado's path of destruction can vary, but it is usually about 8 km long and 10–60 m wide. Although most tornadoes last only a few minutes, they can cause a lot of damage. This is due to their strong spinning winds. The average tornado has wind speeds between 120 and 180 km/h, but rarer, more violent tornadoes can have spinning winds up to 500 km/h. The winds of tornadoes have been known to uproot trees and destroy buildings, as shown in **Figure 21.** Tornadoes are capable of picking up heavy objects, such as mobile homes and cars, and hurling them through the air.

Figure 21 *The tornado that hit Kissimmee, Florida, in 1998 had wind speeds of up to 416 km/h.*

Hurricanes

A **hurricane,** as shown in **Figure 22,** is a large, rotating tropical weather system with wind speeds of at least 119 km/h. Hurricanes are the most powerful storms on Earth. Hurricanes have different names in other parts of the world. In the western Pacific Ocean, they are called *typhoons*. Hurricanes that form over the Indian Ocean are called *cyclones*.

Hurricanes generally form in the area between 5° and 20° north and south latitude over warm, tropical oceans. At higher latitudes, the water is too cold for hurricanes to form. Hurricanes vary in size from 160 km to 1,500 km in diameter, and they can travel for thousands of miles.

Did you know that fish have been known to fall from the sky? Some scientists think the phenomenon of raining fish is caused by waterspouts. A waterspout is a tornado that occurs over water.

Figure 22 Hurricane Fran Photographed from Space

51

RESEARCH

Writing Tell students that creating severe weather takes a lot of energy. Have them research the relationship between energy and storm formation. For example, as a warm air mass rises, energy from water condensation helps fuel hurricanes. The energy released by a typical hurricane in one day is equal to detonating four hundred 20-megaton hydrogen bombs. Challenge students to research these concepts in books, magazines, and the Internet, and compile their findings into a short report.

GROUP ACTIVITY

Have students work in groups to learn about a hurricane of their choosing. Have them find out where the storm formed, its path, the damage it did, and how people recovered from the damage. Ask them to focus on the people involved in the hurricane, from the meteorologists to relief workers. Have each group present the information they gather as a series of simulated newscasts.

GOING FURTHER

Disaster Kit Have groups put together a disaster supply kit that could be used in the event of severe weather. Items that are not easily obtained can be listed on a sheet of paper. Groups can display their kits in class.

Teaching Transparency 174 "A Cross Section of a Hurricane"

Figure 23 *The photo above gives you a bird's-eye view of a hurricane.*

Rain bands

Eye

Eye wall

Formation of a Hurricane A hurricane begins as a group of thunderstorms moving over tropical ocean waters. Winds traveling in two different directions collide, causing the storm to rotate over an area of low pressure. Because of the Coriolis effect, the storm turns counterclockwise in the Northern Hemisphere and clockwise in the Southern Hemisphere.

Hurricanes get their energy from the condensation of water vapor. Once formed, the hurricane is fueled through contact with the warm ocean water. Moisture is added to the warm air by evaporation from the ocean. As the warm, moist air rises, the water vapor condenses, releasing large amounts of energy. The hurricane continues to grow as long as it is over its source of warm, moist air. When the hurricane moves into colder waters or over land, it begins to die because it has lost its source of energy. **Figure 23** and **Figure 24** show two views of a hurricane.

Figure 24 *The view below shows how a hurricane would look if you cut it in half and looked at it from the side. The arrows indicate the flow of air.*

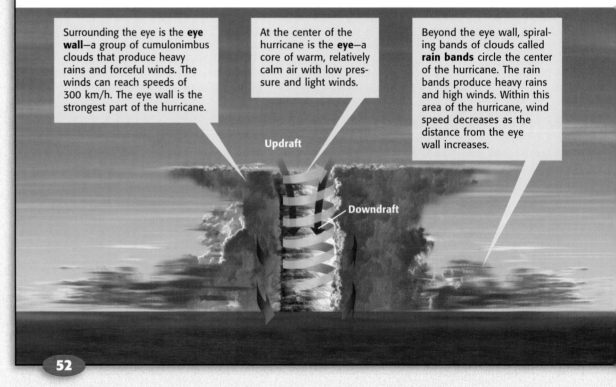

Surrounding the eye is the **eye wall**—a group of cumulonimbus clouds that produce heavy rains and forceful winds. The winds can reach speeds of 300 km/h. The eye wall is the strongest part of the hurricane.

At the center of the hurricane is the **eye**—a core of warm, relatively calm air with low pressure and light winds.

Beyond the eye wall, spiraling bands of clouds called **rain bands** circle the center of the hurricane. The rain bands produce heavy rains and high winds. Within this area of the hurricane, wind speed decreases as the distance from the eye wall increases.

Updraft

Downdraft

52

Science Bloopers

The Seminole Indians of Florida have used their own observations of nature to successfully predict severe weather. In one instance, their observations of plants and animals indicated that a hurricane was approaching. Though the weather bureau predicted the storm would miss the area, the Seminoles evacuated—and were spared the storm's destruction. In another instance, meteorologists were so sure of their predictions that heavy equipment was moved from the area so that it would be available later to help relief efforts. The Seminoles thought otherwise and remained in the area. The hurricane never reached Florida.

Damage Caused by Hurricanes Hurricanes can cause a lot of damage when they move near or onto land. The speed of the steady winds of most hurricanes ranges from 120 km/h to 150 km/h, and they can reach speeds as high as 300 km/h. Hurricane winds can knock down trees and telephone poles and can damage and destroy buildings and homes.

While high winds cause a great deal of damage, most hurricane damage is caused by flooding associated with heavy rains and storm surges. A *storm surge* is a wall of water that builds up over the ocean due to the heavy winds and low atmospheric pressure. The wall of water gets bigger and bigger as it nears the shore, reaching its greatest height when it crashes onto the shore. Depending on the hurricane's strength, a storm surge can be 1 m to 8 m high and 65 km–160 km long. Flooding causes tremendous damage to property and lives when a storm surge moves onto shore, as shown in **Figure 25.**

Astronomy
CONNECTION

The weather on Jupiter is more exciting than that on Earth. Wind speeds reach up to 540 km/h. Storms last for decades, and one—the Great Red Spot of Jupiter—has been swirling around since it was first discovered, in 1664. The Great Red Spot has a diameter of more than one and a half times that of the Earth. It is like a hurricane that has lasted more than 300 years.

Figure 25 *In 1998, the flooding associated with Hurricane Mitch devastated Central America. Whole villages were swept away by the flood waters and mudslides. Thousands of people were killed, and damages were estimated to be more than $5 billion.*

SECTION REVIEW

1. What is lightning?

2. Describe how tornadoes develop. What is the difference between a funnel cloud and a tornado?

3. Why do hurricanes form only over certain areas?

4. **Inferring Relationships** What happens to a hurricane as it moves over land? Why?

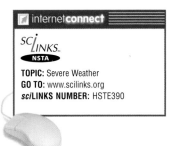
internet**connect**

SC*i*LINKS
NSTA

TOPIC: Severe Weather
GO TO: www.scilinks.org
*sci*LINKS **NUMBER:** HSTE390

Quiz

1. **What is the relationship between lightning and thunder?** (Lightning is an electrical discharge that forms between clouds or between a cloud and the ground. The air around the lightning bolt expands rapidly, producing sound waves that we call thunder.)

2. **Explain why tornadoes often destroy buildings in their path.** (Buildings are often destroyed by the enormous force exerted by tornado winds and by the strong updrafts that accompany them.)

3. **Why don't hurricanes form over land?** (A hurricane gets its energy from enormous volumes of warm, moist air, which are not present over landmasses.)

ALTERNATIVE ASSESSMENT

Concept Mapping Have students make a severe-weather concept map. Tell them that their map should illustrate how thunderstorms, tornadoes, and hurricanes form and what their characteristics are.

Reinforcement Worksheet
"Precipitation Situations"

▼ **Answers to Section Review**

1. Lightning is a large electrical discharge that occurs between two oppositely charged surfaces.

2. A tornado develops when wind traveling in two different directions causes the air in the middle to rotate. The rotating column of air is turned upright by updrafts that begin spinning with it. The rotating air works its way down to the bottom of the cloud and forms a funnel cloud. When the funnel cloud touches the ground, it is called a tornado.

3. Hurricanes form only over warm, tropical oceans because a hurricane requires the energy and moisture from water to fuel it.

4. A hurricane dissipates as it moves over land because it loses its energy source.

Focus

Forecasting the Weather

In this section, students will learn how we use instruments such as thermometers, barometers, weather balloons, and radar to forecast and report the weather. Students will also learn how meteorologists use weather maps to depict the data they gather.

 **Bellringer**

Pose this question to students:

If you did not have the benefit of the weather forecast on the news, radio, or television, how would you forecast the weather? (Answers will vary. Possible answers include observing the sky and noticing the direction and intensity of the winds.)

1 Motivate

DEMONSTRATION

Air Pressure and Barometers
Students have learned that thunderstorms and hurricanes are low-pressure storm systems. Low pressure usually indicates stormy weather, and high pressure usually indicates clear weather. If possible, show students a barometer, and tell them that barometers are still widely used in weather forecasting. Show students how to read a barometer and how to use the moveable pointer to track whether air pressure is increasing or decreasing.
Sheltered English

Terms to Learn

thermometer	wind vane
barometer	anemometer
windsock	isobars

What You'll Do

◆ Describe the different types of instruments used to take weather measurements.
◆ Explain how to interpret a weather map.
◆ Explain why weather maps are useful.

Figure 26 *A liquid thermometer is usually filled with alcohol that is colored red, or mercury, which is silver.*

Directed Reading Worksheet Section 4

Forecasting the Weather

Have you ever left your house in the morning wearing a short-sleeved shirt, only to need a sweater in the afternoon? At some time in your life, you have been caught off guard by the weather. Weather affects how you dress and your daily plans, so it is important that you get accurate weather forecasts. A *weather forecast* is a prediction of weather conditions over the next 3 to 5 days. Meteorologists observe and collect data on current weather conditions in order to provide reliable predictions. In this section you will learn about some of the methods used to collect weather data and how those data are displayed.

Weather Forecasting Technology

In order for meteorologists to accurately forecast the weather, they need to measure various atmospheric conditions, such as air pressure, humidity, precipitation, temperature, wind speed, and wind direction. Meteorologists use special instruments to collect data on weather conditions both near and far above the Earth's surface. You have already learned about two tools that meteorologists use near the Earth's surface—psychrometers, which are used to measure relative humidity, and rain gauges, which are used to measure precipitation. Read on to learn about other methods meteorologists use to collect data.

Measuring Air Temperature A **thermometer** is a tool used to measure air temperature. A common type of thermometer uses a liquid sealed in a narrow glass tube, as shown in **Figure 26.** When air temperature increases, the liquid expands and moves up the glass tube. As air temperature decreases, the liquid shrinks and moves down the tube.

Air temperature is measured in both degrees Celsius and degrees Fahrenheit. In the United States, television weather forecasters generally report air temperature in degrees Fahrenheit.

MISCONCEPTION
ALERT

Students may think that the lowest and highest temperatures occur in the middle of the night and in the middle of the day. Actually, the lowest temperatures usually occur around sunrise because the Earth's surface has radiated thermal energy all night. The highest temperatures usually occur in the late afternoon.

Measuring Air Pressure A **barometer** is an instrument used to measure air pressure. The mercurial barometer provides the most accurate method of measuring air pressure. A mercurial barometer consists of a glass tube sealed at one end that is placed in a container full of mercury. The air pressure pushes on the mercury inside the container, causing the mercury to move up the glass tube. The greater the air pressure is, the higher the mercury will rise.

Measuring Wind Direction Wind direction can be measured using a **windsock** or a **wind vane**. A windsock, as shown in **Figure 27,** is a cone-shaped cloth bag open at both ends. The wind enters through the wide end and leaves through the narrow end. Therefore, the wide end points into the wind.

A wind vane is shaped like an arrow with a large tail and is attached to a pole. The wind pushes the tail of the wind vane, spinning it on the pole until the arrow points into the wind.

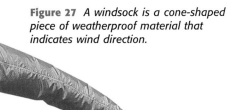

Figure 27 *A windsock is a cone-shaped piece of weatherproof material that indicates wind direction.*

Measuring Wind Speed Wind speed is measured by a device called an **anemometer.** An anemometer, as shown in **Figure 28,** consists of three or four cups connected by spokes to a pole. The wind pushes on the hollow sides of the cups, causing them to rotate on the pole. The motion sends a weak electrical current that is measured and displayed on a dial.

Measuring Weather in the Upper Atmosphere You have learned how weather conditions are recorded near the Earth's surface. But in order for meteorologists to better understand weather patterns, they must collect data from higher altitudes. Studying weather at higher altitudes requires the use of more-sophisticated equipment.

Figure 28 *The faster the wind speed is, the faster the cups of the anemometer spin.*

CONNECT TO
LIFE SCIENCE

Middle ear barotrauma is an earache caused by a difference in pressure between the air and a person's middle ear. Although airplane cabins are pressurized, passengers still feel the pressure decrease as the plane climbs and increase as it descends. The trauma occurs when the Eustachian tube, a passageway between the middle ear and the throat, fails to open wide enough to equalize the pressure. Chewing gum, yawning, or simply swallowing often alleviates the condition.

Before sophisticated weather forecasts, people learned to carefully observe the world around them for evidence of changing weather. These clues can be found everywhere. Have students research and test these and other observations.

- Birds fly higher when fair weather is coming. (They fly high to avoid the increased air resistance of a high pressure air mass.)
- Heavy dew condenses early in fair night air. If there is little or no dew, the chance for rain is good.
- Leaves turn over in nonprevailing storm winds.
- Halos form around the sun or moon as light shines through ice particles in the clouds of an advancing rainstorm.
- As a pre-rain low pressure front moves in, odors trapped in objects by high pressure air masses are suddenly released.
- Ants travel in lines when rain is coming and scatter when the weather is clear.
- Robins sing high in fair weather and sing low if rain is approaching.
- Flying insects swarm before a rain; they bite the most when the air is moist.
- Clouds lower as a low pressure system approaches. This signals that a storm is coming.
- Swallows and bats fly lowest when air pressure decreases before a storm. Their sensitive ears are more comfortable when they are flying close to the ground (where air pressure is highest).
- At 15°C, rhododendron leaves stand upright, at 4°C they droop, at –1°C they curl, and at –6°C they turn black.

GROUP ACTIVITY

Doppler radar is a type of radar that uses the Doppler effect to determine the direction and speed of weather systems. Radar, an acronym for *radio detection and ranging*, is a device that determines the speed and location of weather systems or moving objects by bouncing radio waves off them. The Doppler effect is the shift in wave frequency detected by an observer due to the motion of the wave source relative to the observer. For example, an ambulance siren may emit sound waves at a uniform rate. But as the ambulance passes you, you may notice that the pitch drops. The reason is that the sound waves seem to be compressed as the ambulance approaches you and seem to spread out as it speeds away. Doppler radar bounces radio waves off weather systems. The reflected waves are used to determine if the system is moving toward or away from the radar source and at what speed it is moving. Challenge students to illustrate the Doppler effect so that the concept makes sense to them, perhaps by making a poster or demonstrating the effect using water waves.

MATH and MORE

Have students use the formulas below to convert 32°F, 72°F, and 5°F into degrees Celsius. Then convert 100°C, 45°C, and 21°C into degrees Fahrenheit.

$C = (F - 32) \times \frac{5}{9}$

$F = (C \times \frac{9}{5}) + 32$

(0°C, 22.2°C, −15°C, 212°F, 113°F, 69.8°F)

Math Skills Worksheet
"Using Temperature Scales"

Figure 29 *Weather balloons carry radio transmitters that send measurements to stations on the ground.*

Activity

Throughout history, people have predicted approaching weather by interpreting natural signs. Animals and plants are usually more sensitive to changes in the atmosphere, such as air pressure, humidity, and temperature, than humans. To find out more about natural signs, research this topic at the library or on the Internet. Try searching using key words and phrases such as "weather and animals" or "weather and plants." Write a short paper on your findings to share with the class.

TRY at HOME

Eyes in the Sky Weather balloons carry electronic equipment that can measure weather conditions as high as 30 km above the Earth's surface. Weather balloons, such as the one in **Figure 29,** carry equipment that measures temperature, air pressure, and relative humidity.

Radar is used to find the location, movement, and intensity of precipitation. It can also detect what form of precipitation a weather system is carrying. You might be familiar with a type of radar called Doppler radar. **Figure 30** shows how Doppler radar is used to track precipitation.

Figure 30 *Using Doppler radar, meteorologists can predict a tornado up to 20 minutes before it touches the ground.*

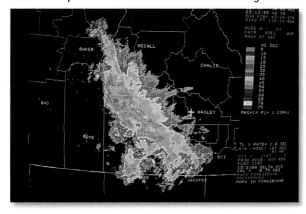

Weather satellites orbiting the Earth provide the images of the swirling clouds you can see on television weather reports. Satellites can measure wind speeds, humidity, and the temperatures at various altitudes.

Weather Maps

As you have learned, meteorologists base their forecasts on information gathered from many sources. In the United States, the National Weather Service (NWS) and the National Oceanic and Atmospheric Administration (NOAA) collect and analyze weather data. The NWS produces weather maps based on information gathered from about 1,000 weather stations across the United States. On these maps, each station is represented by a station model. A *station model,* as shown in **Figure 31,** is a small circle, which shows the location of the weather station, with a set of symbols and numbers surrounding it, which represent the weather data.

CROSS-DISCIPLINARY FOCUS

Language Arts Have students find poems about weather and share them with the class. Students could also write their own poem about their favorite and least favorite kinds of weather. The poems should demonstrate some knowledge about the cause of the weather they describe.

IS THAT A FACT!

Bats use the Doppler effect to locate prey and to navigate. Bats emit high-frequency sounds that bounce off objects. If the objects are moving, the wave frequency changes. If the frequency doesn't change, then the bat knows the object is stationary.

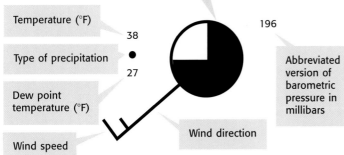

Temperature (°F)

Type of precipitation

Dew point temperature (°F)

Wind speed

Amount of cloud cover

Wind direction

Abbreviated version of barometric pressure in millibars

Figure 31 *Weather conditions at a station are represented by symbols.*

38

27

196

Under Pressure Weather maps also include lines called isobars. Isobars are similar to contour lines on a topographical map, except **isobars** are lines that connect points of equal air pressure rather than equal elevation. Isobar lines that form closed circles represent areas of high or low pressure. These areas are usually marked on a map with a capital *H* or *L*. Fronts are also labeled on weather maps. Weather maps, like the one shown in **Figure 32,** provide useful information for making accurate weather forecasts.

Figure 32 *Can you identify the different fronts on the weather map?*

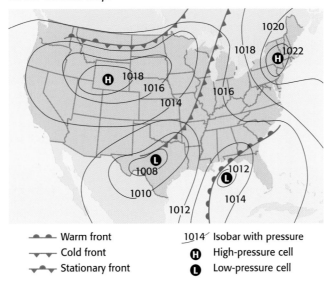

🔺🔺 Warm front
▽▽ Cold front
🔺▽ Stationary front

<u>1014</u> Isobar with pressure
Ⓗ High-pressure cell
Ⓛ Low-pressure cell

SECTION REVIEW

1. What are three methods meteorologists use to collect weather data?

2. What are weather maps based on?

3. What does a station model represent?

4. **Inferring Conclusions** Why would a meteorologist compare a new weather map with one 24 hours old?

📶 internet**connect**

SC*i*LINKS™
NSTA

TOPIC: Forcasting the Weather
GO TO: www.scilinks.org
*sci*LINKS NUMBER: HSTE395

Quiz

1. Would water be a useful fluid to use in a thermometer? Explain. (No, water would not be a good thermometer fluid because it expands when it freezes.)

2. What advantage do weather satellites have over ground-based weather stations? (Satellites can gather weather data from much higher altitudes than land-based instruments can.)

3. Why are so many station models used to gather weather data in the United States? (Because the country is so large, and Earth's atmosphere is constantly changing, we need data from many stations to make accurate forecasts.)

ALTERNATIVE ASSESSMENT

Have students use the weather report from their local newspaper over a 1-week period to construct a picture of local weather conditions. Then tell them to analyze their findings by applying what they have learned in this chapter.

PG 104

Watching the Weather

▼ *Answers to Section Review*

1. Answers will vary. Sample answer: weather balloons, Doppler radar, and weather satellites

2. Weather maps are based on weather data gathered from weather stations across the United States.

3. A station model represents the location of the weather station and the weather data collected there.

4. Answers may vary. Sample answer: Meteorologists would compare a new weather map with one 24 hours old to see how fast a front is moving.

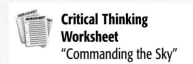

Critical Thinking Worksheet "Commanding the Sky"

Design Your Own Lab

Gone with the Wind
Teacher's Notes

Time Required

One 45-minute class period

Lab Ratings

EASY ————————————> HARD

TEACHER PREP ▲▲
STUDENT SET-UP ▲▲
CONCEPT LEVEL ▲▲
CLEAN UP ▲

MATERIALS

This activity works best in groups of 2–3 students.

Safety Caution

Remind students to review all safety cautions and icons before beginning this lab activity.

Preparation Notes

You might want to watch your local weather station in order to schedule this experiment on a windy day. Use a magnetic compass to find magnetic north. Then use masking tape or chalk to mark the sidewalk or parking lot with an arrow pointing toward magnetic north. Before the activity, ask students if they have ever seen a weather vane. Also have them list several reasons why knowing the wind direction might be helpful.

Gone with the Wind

Pilots at the Fly Away Airport need your help—fast! Last night, lightning destroyed the orange windsock. This windsock helped pilots measure which direction the wind was blowing. But now the windsock is gone with the wind, and an incoming airplane needs to land. The pilot must know which direction the wind is blowing and is counting on you to make a device that can measure wind direction.

MATERIALS

- paper plate
- drawing compass
- metric ruler
- protractor
- index card
- scissors
- stapler
- straight plastic straw
- sharpened pencil
- thumbtack or pushpin
- magnetic compass
- small rock

Ask a Question

1 How can I measure wind direction?

Conduct an Experiment

2 Find the center of the plate by tracing around its edge with a drawing compass. The pointed end of the compass should poke a small hole in the center of the plate.

3 Use a ruler to draw a line across the center of the plate.

4 Use a protractor to help you draw a second line through the center of the plate. This new line should be at a 90° angle to the line you drew in step 3.

5 Moving clockwise, label each line *N*, *E*, *S*, and *W*.

6 Use a protractor to help you draw two more lines through the center of the plate. These lines should be at a 45° angle to the lines you drew in steps 3 and 4.

58

CLASSROOM
TESTED & APPROVED

Walter Woolbaugh
Manhattan School System
Manhattan, Montana

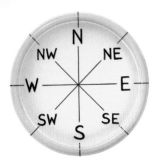

7 Moving clockwise from *N*, label these new lines *NE*, *SE*, *SW*, and *NW*. The plate now resembles the face of a magnetic compass. This will be the base of your wind-direction indicator. It will help you read the direction of the wind at a glance.

8 Measure and mark a 5 cm × 5 cm square on an index card. Cut the square out of the card. Fold the square in half to form a triangle.

9 Staple an open edge of the triangle to the straw so that one point of the triangle touches the end of the straw.

10 Hold the pencil at a 90° angle to the straw. The eraser should touch the balance point of the straw. Push a thumbtack or pushpin through the straw and into the eraser. The straw should spin without falling off.

11 Find a suitable area outside to measure the wind direction. The area should be clear of trees and buildings.

12 Press the sharpened end of the pencil through the center hole of the plate and into the ground. The labels on your paper plate should be facing the sky, as shown below.

13 Use a compass to find magnetic north. Rotate the plate so that the *N* on the plate points north. Place a small rock on top of the plate so that it does not turn.

14 You have just constructed a wind vane. Watch the straw as it rotates. The triangle will point in the direction the wind is blowing.

Analyze the Results

15 From which direction is the wind coming?

16 In which direction is the wind blowing?

Draw Conclusions

17 Would this be an effective way for pilots to measure wind direction? Why or why not?

18 What improvements would you suggest to Fly Away Airport to measure wind direction more accurately?

Answers

15. Answers will vary.

16. Answers will vary.

17. Answers will vary. Accept all reasonable responses.

18. Answers will vary. Accept all reasonable responses.

Datasheets for LabBook

Science Skills Worksheet "Using Models to Communicate"

Chapter Highlights

Chapter Highlights

VOCABULARY DEFINITIONS

SECTION 1

weather the condition of the atmosphere at a particular time and place

water cycle the continuous movement of water from water sources into the air, onto land, into and over the ground, and back to the water sources; a cycle that links all of the Earth's solid, liquid, and gaseous water together

humidity the amount of water vapor or moisture in the air

relative humidity the amount of moisture the air contains compared with the maximum amount it can hold at a particular temperature

condensation the change of state from a gas to a liquid

dew point the temperature to which air must cool to be completely saturated

cloud a collection of millions of tiny water droplets or ice crystals

precipitation solid or liquid water that falls from the air to the Earth

SECTION 2

air mass a large body of air that has similar temperature and moisture throughout

front the boundary that forms between two different air masses

 Vocabulary Review Worksheet

SECTION 1

Vocabulary
- **weather** (p. 36)
- **water cycle** (p. 36)
- **humidity** (p. 37)
- **relative humidity** (p. 37)
- **condensation** (p. 39)
- **dew point** (p. 39)
- **cloud** (p. 40)
- **precipitation** (p. 42)

Section Notes

- Water is continuously moving and changing state as it moves through the water cycle.

- Humidity is the amount of water vapor or moisture in the air. Relative humidity is the amount of moisture the air contains compared with the maximum amount it can hold at a particular temperature.

- Water droplets form because of condensation.

- Dew point is the temperature to which air must cool to be saturated.

- Condensation occurs when the air next to a surface cools to below its dew point.

- Clouds are formed from condensation on dust and other particles above the ground.

- There are three major cloud forms—cumulus, stratus, and cirrus.

- There are four major forms of precipitation—rain, snow, sleet, and hail.

Labs
Let It Snow! (p. 107)

SECTION 2

Vocabulary
- **air mass** (p. 44)
- **front** (p. 46)

Section Notes

- Air masses form over source regions. An air mass has similar temperature and moisture content throughout.

- Four major types of air masses influence weather in the United States—maritime polar, maritime tropical, continental polar, continental tropical.

- A front is a boundary between contrasting air masses.

- There are four types of fronts—cold fronts, warm fronts, occluded fronts, and stationary fronts.

- Specific types of weather are associated with each front.

☑ Skills Check

Math Concepts

RELATIVE HUMIDITY Relative humidity is the amount of moisture the air is holding compared with the amount it can hold at a particular temperature. The relative humidity of air that is holding all the water it can at a given temperature is 100 percent, meaning it is saturated. You can calculate relative humidity with the following equation:

$$\frac{(present)\ g/m^3}{(saturated)\ g/m^3} \times 100 = relative\ humidity$$

Visual Understanding

HURRICANE HORSEPOWER Hurricanes are the most powerful storms on Earth. A cross-sectional view helps you identify the different parts of a hurricane. The diagram on page 52 shows a side view of a hurricane.

Lab and Activity Highlights

Gone with the Wind PG 58

Watching the Weather PG 104

Let It Snow! PG 107

 Datasheets for LabBook (blackline masters for these labs)

60

SECTION 3

Vocabulary

thunderstorm *(p. 48)*

lightning *(p. 49)*

thunder *(p. 49)*

tornado *(p. 50)*

hurricane *(p. 51)*

Section Notes

- Severe weather is weather that can cause property damage and even death.

- Thunderstorms are small, intense storm systems that produce lightning, thunder, strong winds, and heavy rain.

- Lightning is a large electrical discharge that occurs between two oppositely charged surfaces.

- Thunder is the sound that results from the expansion of air along a lightning strike.

- A tornado is a rotating funnel cloud that touches the ground.

- Hurricanes are large, rotating, tropical weather systems that form over the tropical oceans.

SECTION 4

Vocabulary

thermometer *(p. 54)*

barometer *(p. 55)*

windsock *(p. 55)*

wind vane *(p. 55)*

anemometer *(p. 55)*

isobars *(p. 57)*

Section Notes

- Weather balloons, radar, and weather satellites take weather measurements at high altitudes.

- Meteorologists present weather data gathered from stations as station models on weather maps.

Labs

Watching the Weather *(p. 104)*

VOCABULARY DEFINITIONS, continued

SECTION 3

thunderstorm a small, intense weather system that produces strong winds, heavy rain, lightning, and thunder

lightning the large electrical discharge that occurs between two oppositely charged surfaces

thunder the sound that results from the rapid expansion of air along a lightning strike

tornado a small, rotating column of air that has high wind speeds and low central pressure and that touches the ground

hurricane a large, rotating tropical weather system with wind speeds of at least 119 km/h

SECTION 4

thermometer a tool used to measure air temperature

barometer an instrument used to measure air pressure

windsock a device used to measure wind direction

wind vane a device used to measure wind direction

anemometer a device used to measure wind speed

isobars lines that connect points of equal air pressure

 Blackline masters of these Chapter Highlights can be found in the **Study Guide.**

---

internet connect

GO TO: go.hrw.com

Visit the **HRW** Web site for a variety of learning tools related to this chapter. Just type in the keyword:

KEYWORD: HSTWEA

SCI **LINKS**ₛₘ
N S T A

GO TO: www.scilinks.org

Visit the **National Science Teachers Association** on-line Web site for Internet resources related to this chapter. Just type in the *sci*LINKS number for more information about the topic:

TOPIC: Collecting Weather Data	*sci*LINKS NUMBER: HSTE380
TOPIC: Air Masses and Fronts	*sci*LINKS NUMBER: HSTE385
TOPIC: Severe Weather	*sci*LINKS NUMBER: HSTE390
TOPIC: Forecasting the Weather	*sci*LINKS NUMBER: HSTE395

61

Lab and Activity Highlights

LabBank

 Whiz-Bang Demonstrations
- It's Raining Again
- When Air Bags Collide

Calculator-Based Labs, Relative Humidity

Inquiry Labs, When Disaster Strikes

 EcoLabs & Field Activities, Rain Maker or Rain Faker?

Long-Term Projects & Research Ideas, A Storm on the Horizon

Chapter Review
Answers

USING VOCABULARY

1. Relative humidity is the amount of water vapor the air contains relative to the maximum amount it can hold at a given temperature. Dew point is the temperature to which air must cool to be saturated.

2. Condensation is the process by which a gas changes state to become a liquid. Precipitation is liquid or solid water that falls from the atmosphere to the Earth.

3. An air mass is a large body of air that has the same moisture and temperature throughout. A front is the boundary that forms where two different air masses meet.

4. Lightning is a large electrical discharge that occurs between two oppositely charged surfaces. Thunder is the sound that results from the rapid expansion of air along a lightning strike.

5. A tornado is a small, rotating column of air with high wind speed that touches the ground. A hurricane is a large, rotating tropical weather system with wind speeds equal to or greater than 119 km/h.

6. A barometer measures air pressure. An anemometer measures wind speed.

UNDERSTANDING CONCEPTS

Multiple Choice

7. c
8. d
9. c
10. d
11. b
12. d
13. a
14. c
15. b
16. c

Chapter Review

USING VOCABULARY

Explain the difference between the following sets of words:

1. relative humidity/dew point

2. condensation/precipitation

3. air mass/front

4. lightning/thunder

5. tornado/hurricane

6. barometer/anemometer

UNDERSTANDING CONCEPTS

Multiple Choice

7. The process of liquid water changing to gas is called
   a. precipitation.
   b. condensation.
   c. evaporation.
   d. water vapor.

8. What is the relative humidity of air at its dew-point temperature?
   a. 0 percent
   b. 50 percent
   c. 75 percent
   d. 100 percent

9. Which of the following is not a type of condensation?
   a. fog
   b. cloud
   c. snow
   d. dew

10. High clouds made of ice crystals are called ___?___ clouds.
    a. stratus
    b. cumulus
    c. nimbostratus
    d. cirrus

11. Large thunderhead clouds that produce precipitation are called ___?___ clouds.
    a. nimbostratus
    b. cumulonimbus
    c. cumulus
    d. stratus

12. Strong updrafts within a thunderhead can produce
    a. snow.
    b. rain.
    c. sleet.
    d. hail.

13. A maritime tropical air mass contains
    a. warm, wet air.
    b. cold, moist air.
    c. warm, dry air.
    d. cold, dry air.

14. A front that forms when a warm air mass is trapped between cold air masses and forced to rise is called a(n)
    a. stationary front.
    b. warm front.
    c. occluded front.
    d. cold front.

15. A severe storm that forms as a rapidly rotating funnel cloud is called a
    a. hurricane.
    b. tornado.
    c. typhoon.
    d. thunderstorm.

16. The lines on a weather map connecting points of equal atmospheric pressure are called
    a. contour lines.
    b. highs.
    c. isobars.
    d. lows.

Short Answer

17. Explain the relationship between condensation and the dew point.

18. Describe the conditions along a stationary front.

19. What are the characteristics of an air mass that forms over the Gulf of Mexico?

20. Explain how a hurricane develops.

Concept Mapping

21. Use the following terms to create a concept map: evaporation, relative humidity, water vapor, dew, psychrometer, clouds, fog.

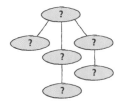

CRITICAL THINKING AND PROBLEM SOLVING

Write one or two sentences to answer the following questions:

22. If both the air temperature and the amount of water vapor in the air change, is it possible for the relative humidity to stay the same? Explain.

23. What can you assume about the amount of water vapor in the air if there is no difference between the wet- and dry-bulb readings of a psychrometer?

24. List the major similarities and differences between hurricanes and tornadoes.

MATH IN SCIENCE

You always see lightning before you hear thunder. That's because light travels at about 300,000,000 m/s, while sound travels only 330 m/s. One way you can determine how close you are to the thunderstorm is by counting how many seconds there are between the lightning and thunder. Usually, it takes thunder about 3 seconds to cover 1 km. Answer the following questions based on this estimate.

25. If you hear thunder 12 seconds after you see the flash of lightning, how far away is the thunderstorm?

26. If you hear thunder 36 seconds after you see the flash of lightning, how far away is the thunderstorm?

INTERPRETING GRAPHICS

Use the weather map below to answer the questions that follow.

Seattle

Boston

Chicago

San Diego

Tulsa

◄▲◄▲ Warm front

▼▼ Cold front

▼▲▼▲ Stationary front

27. Where are thunderstorms most likely to occur? Explain your answer.

28. What are the weather conditions like in Tulsa, Oklahoma? Explain your answer.

Reading Check-up

Take a minute to review your answers to the Pre-Reading Questions found at the bottom of page 84. Have your answers changed? If necessary, revise your answers based on what you have learned since you began this chapter.

63

Concept Mapping Transparency 16

Blackline masters of this Chapter Review can be found in the **Study Guide.**

UNDERSTANDING CONCEPTS

Short Answer

17. The air must cool to below its dew point before condensation can occur.
18. Stationary fronts generally bring drizzly precipitation. After the front passes, the weather is generally clear and warm.
19. An air mass that forms over the Gulf of Mexico is warm and wet.
20. A hurricane begins as a group of thunderstorms moving over tropical ocean waters. Winds traveling in two different directions collide, causing the storm to rotate over an area of low pressure. The hurricane is fueled by the condensation of water vapor.

Concept Mapping

21. An answer to this exercise can be found at the front of this book.

CRITICAL THINKING AND PROBLEM SOLVING

22. Yes; for example, if both the air temperature and water vapor increased, the relative humidity might remain the same.
23. The air is saturated with water.
24. Answers may vary. Sample answer: Both begin as a result of thunderstorms and are centered around low pressure. Hurricanes occur over water, and tornadoes generally occur over land.

MATH IN SCIENCE

25. (12 s ÷ 3 s) × 1 km = 4 km
26. (36 s ÷ 3 s) × 1 km = 12 km

INTERPRETING GRAPHICS

27. Thunderstorms are most likely to occur in Chicago because a cold front is approaching.
28. Tulsa is experiencing a stationary front. It is probably receiving drizzly precipitation.

CAREERS

CAREERS
Meteorologist– Cristy Mitchell

Teaching Strategy

Tell students to imagine that they are meteorologists studying another planet in the solar system. Ask:

As a meteorologist, what features would you look for to get information about the climate and the weather? (Students should recognize that climates are strongly influenced by major geographic features. A planet's rotation affects prevailing wind patterns. Other features to look for include mountains, deserts, and large bodies of water.)

Discussion

1. Weather stations operate around the clock, 7 days a week. Do you think you would enjoy the night work and rotating shifts that are part of a meteorologist's job? (Some students may think this is exciting, while others may prefer a 9-to-5 job.)

2. There is an old saying: "You can talk about the weather, but you can't do anything about it." How might this relate to a meteorologist's job? (Although scientists like Cristy Mitchell observe the powerful forces of nature, they cannot do anything to stop them. However, issuing accurate weather warnings can save people's lives.)

METEOROLOGIST

Predicting floods, observing a tornado develop inside a storm, watching the growth of a hurricane, and issuing flood warnings are all in a day's work for **Cristy Mitchell.** As a meteorologist for the National Weather Service, Mitchell spends each working day observing the powerful forces of nature.

In addition to using computers, Mitchell also uses radar and satellite imagery to show regional and national weather. Meteorologists also use computerized models of the world's atmosphere to help forecast the weather.

Find Out for Yourself

▶ Use the library or the Internet to find information about hurricanes, tornadoes, or thunderstorms. How do meteorologists define these storms? What trends in air pressure, temperature, and humidity do meteorologists use to forecast storms?

When asked what made her job interesting, Mitchell replied, "There's nothing like the adrenaline rush you get when you see a tornado coming! I would say that witnessing the powerful forces of nature is what really makes my job interesting."

Meteorology is the study of natural forces in Earth's atmosphere. Perhaps the most familiar field of meteorology is weather forecasting. However, meteorology is also used in air-pollution control, agricultural planning, and air and sea transportation, and criminal and civil investigations. Meteorologists also study trends in Earth's climate, such as global warming and ozone depletion.

Collecting the Data

Meteorologists collect data on air pressure, temperature, humidity, and wind velocity. By applying what they know about the physical properties of the atmosphere and analyzing the mathematical relationships in the data, they are able to forecast the weather.

Meteorologists use a variety of tools, such as computers and satellites, to collect the data they need to make accurate weather forecasts. Mitchell explained, "The computer is an invaluable tool for me. Through it, I receive maps and detailed information, including temperature, wind speed, air pressure, lightning activity, and general sky conditions for a specific region."

▲ *This photograph of Hurricane Elena was taken from the space shuttle* Discovery *in September 1985.*

64

Answers to Find Out for Yourself

Students' answers will vary depending upon which topic they choose to research.

Science Fiction

"All Summer in a Day"

by Ray Bradbury

It is raining, just like it has been for 7 long years. That is 2,555 days of nonstop rain. For the men, women, and children who came to build a civilization on Venus, constant rain is a fact of life. But there is one special day—a day when it stops raining and the sun shines gloriously. This day comes about only once every 7 years. And today is that day!

At school the students have been looking forward to this day for weeks. In one class they've read about how the sun is like a lemon, and how hot it is. They've written stories and poems about what it might be like to see the sun.

And now that the day has finally arrived, all of the children in that class are peering through the window, searching for the sun. The children are 9 years old, and all of them but Margot have lived on Venus all their lives. None of them remember the day 7 years ago when the rain stopped. They only recall stories about the sunshine, and now they just can't wait to see it for themselves!

But Margot is different. She longs to see the sun even more than the others. The reason makes the other kids jealous. And jealous kids can be cruel. . . .

What happens to Margot? Find out for yourself by reading Ray Bradbury's "All Summer in a Day" in the *Holt Anthology of Science Fiction.*

65

Further Reading Students can check out some of Ray Bradbury's other classic stories in the following collections. Or they can visit the library to scan the wide range of Bradbury's publications.

The Veldt, Creative Education, Inc., 1987

The Foghorn, Creative Education, Inc., 1987

S is for Space, Doubleday, 1966

SCIENCE FICTION
"All Summer in a Day"
by Ray Bradbury

Often, a memory can be sustaining; other times, it might be crippling. A priceless experience teaches Margot's classmates a lesson in understanding the power of memory.

Teaching Strategy

Reading Level This is a relatively short story that should not be difficult for the average student to read and comprehend.

Background

About the Author Ray Bradbury is one of the world's most celebrated writers. He was born in the small town of Waukegan, Illinois, in 1920. He moved from place to place as a young boy while his father looked for steady work. Eventually, Bradbury and his family ended up in Los Angeles. There he began a writing career that has spanned over 60 years!

Bradbury has earned top honors in the field of literature, including the World Fantasy Award for lifetime work and the Grand Master Award from Science Fiction Writers of America. An unusual honor came when an astronaut named a crater on the moon Dandelion Crater after Ray Bradbury's novel, *Dandelion Wine.*

Chapter Organizer

CHAPTER ORGANIZATION	TIME MINUTES	OBJECTIVES	LABS, INVESTIGATIONS, AND DEMONSTRATIONS
Chapter Opener pp. 66–67	45	National Standards: SAI 1	**Start-Up Activity,** What's Your Angle? p. 67
Section 1 **What Is Climate?**	90	▶ Explain the difference between weather and climate. ▶ Identify the factors that determine climates. SAI 1, SPSP 1, 3, ES 1j, 3d	**QuickLab,** A Cool Breeze, p. 71 **Whiz-Bang Demonstrations,** How Humid Is It? **Calculator-Based Labs,** What Causes the Seasons?
Section 2 **Climates of the World**	90	▶ Locate and describe the three major climate zones. ▶ Describe the different biomes found in each climate zone. HNS 1, 3; Labs UCP 2, 3, SAI 1, ST 1	**Demonstration,** Mock Permafrost, p. 81 in ATE **Discovery Lab,** For the Birds, p. 109 **Datasheets for LabBook,** For the Birds **Skill Builder,** Biome Business, p. 88 **Datasheets for LabBook,** Biome Business **Calculator-Based Labs,** Heating of Land and Water
Section 3 **Changes in Climate**	135	▶ Describe how the Earth's climate has changed over time. ▶ Summarize the different theories that attempt to explain why the Earth's climate has changed. ▶ Explain the greenhouse effect and its role in global warming. UCP 2, 3, SAI 1, 2, ST 2, SPSP 3–5, HNS 1, ES 1k, 2a; Labs UCP 2, 3, SAI 1	**Demonstration,** The Greenhouse Effect, p. 83 in ATE **Skill Builder,** Global Impact, p. 108 **Datasheets for LabBook,** Global Impact **Long-Term Projects & Research Ideas,** Sun-Starved in Fairbanks

---

See page **T23** *for a complete correlation of this book with the*

NATIONAL SCIENCE EDUCATION STANDARDS.

TECHNOLOGY RESOURCES

 Guided Reading Audio CD
English or Spanish, Chapter 3

 One-Stop Planner CD-ROM with Test Generator

 Science Discovery Videodiscs
Image and Activity Bank with Lesson Plans:
Exploring Antarctica, Global Warming

 CNN. Eye on the Environment, A Climate Conference, Segment 27

Scientists in Action, Ice Age Discoveries, Segment 23

CLASSROOM WORKSHEETS, TRANSPARENCIES, AND RESOURCES	SCIENCE INTEGRATION AND CONNECTIONS	REVIEW AND ASSESSMENT
Directed Reading Worksheet **Science Puzzlers, Twisters & Teasers**		
Directed Reading Worksheet, Section 1 **Transparency 175,** Seasons, Latitude, and the Tilt of the Earth **Transparency 176,** Basic Properties of Air **Transparency 177,** An Example of the Rain Shadow Effect	**Multicultural Connection,** p. 68 in ATE **Connect to Geography,** p. 69 in ATE **Math and More,** p. 69 in ATE **Multicultural Connection,** p. 71 in ATE **Across the Sciences:** Blame "The Child," p. 94	**Self-Check,** p. 70 **Homework,** p. 71 in ATE **Section Review,** p. 73 **Quiz,** p. 73 in ATE **Alternative Assessment,** p. 73 in ATE
Transparency 178, Climate Zones of the Earth **Transparency 178,** The Earth's Land Biomes **Directed Reading Worksheet,** Section 2 **Transparency 52,** Transpiration **Math Skills for Science Worksheet,** Rain-Forest Math **Science Skills Worksheet,** Finding Useful Sources **Reinforcement Worksheet,** A Tale of Three Climates	**Math and More,** p. 75 in ATE **Connect to Environmental Science,** p. 75 in ATE **Biology Connection,** p. 76 **Real-World Connection,** pp. 76, 77 in ATE **Connect to Environmental Science,** pp. 78, 79 in ATE **Environment Connection,** p. 81 **Physics Connection,** p. 82	**Self-Check,** p. 76 **Homework,** pp. 76, 79, 80 in ATE **Section Review,** p. 77 **Section Review,** p. 82 **Quiz,** p. 82 in ATE **Alternative Assessment,** p. 82 in ATE
Directed Reading Worksheet, Section 3 **Transparency 179,** The Milankovitch Theory of the Causes of the Ice Ages **Critical Thinking Worksheet,** Cyberspace Heats Up **Science Skills Worksheet,** Understanding Bias	**Real-World Connection,** p. 85 in ATE **Cross-Disciplinary Focus,** p. 85 in ATE **Multicultural Connection,** p. 85 in ATE **MathBreak,** The Ride to School, p. 86 **Connect to Life Science,** p. 86 in ATE **Apply,** p. 87 **Science, Technology, and Society:** Some Say Fire, Some Say Ice..., p. 95	**Self-Check,** p. 84 **Section Review,** p. 87 **Quiz,** p. 87 in ATE **Alternative Assessment,** p. 87 in ATE

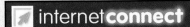

Holt, Rinehart and Winston On-line Resources

go.hrw.com

For worksheets and other teaching aids related to this chapter, visit the HRW Web site and type in the keyword: **HSTCLM**

National Science Teachers Association

www.scilinks.org

Encourage students to use the *sci*LINKS numbers listed in the internet connect boxes to access information and resources on the **NSTA** Web site.

END-OF-CHAPTER REVIEW AND ASSESSMENT

Chapter Review in Study Guide
Vocabulary and Notes in Study Guide
Chapter Tests with Performance-Based Assessment, Chapter 3 Test
Chapter Tests with Performance-Based Assessment, Performance-Based Assessment 3
Concept Mapping Transparency 17

Chapter Resources & Worksheets

Visual Resources

TEACHING TRANSPARENCIES

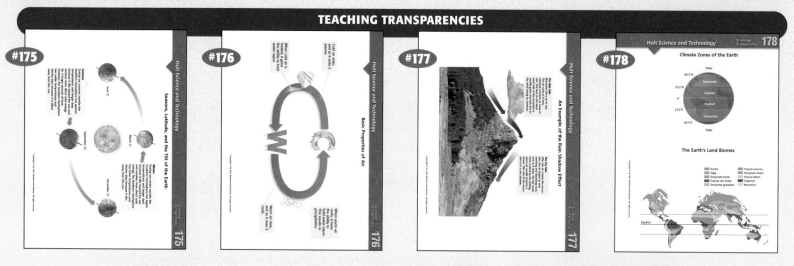

#175 Holt Science and Technology — Seasons, Latitude, and the Tilt of the Earth — *Teaching Transparency* **175**

#176 Holt Science and Technology — Basic Properties of Air — *Teaching Transparency* **176**

#177 Holt Science and Technology — An Example of the Rain Shadow Effect — *Teaching Transparency* **177**

#178 Holt Science and Technology — *Teaching Transparency* **178**
Climate Zones of the Earth
The Earth's Land Biomes

TEACHING TRANSPARENCIES

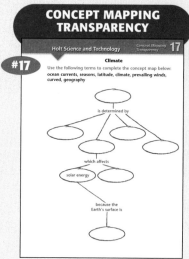

#179 Holt Science and Technology — The Milankovitch Theory of the Causes of the Ice Ages — *Teaching Transparency* **179**

#52 Holt Science and Technology — Transpiration — **52**

LINK TO LIFE SCIENCE

CONCEPT MAPPING TRANSPARENCY

#17 Holt Science and Technology — *Concept Mapping Transparency* **17**
Climate
Use the following terms to complete the concept map below:
ocean currents, seasons, latitude, climate, prevailing winds, curved, geography

is determined by

which affects

solar energy

because the
Earth's surface is

Meeting Individual Needs

DIRECTED READING

#3 DIRECTED READING WORKSHEET
Climate

Chapter Introduction
As you begin this chapter, answer the following.

1. Read the title of the chapter. List three things that you already know about this subject.

2. Write two questions about this subject that you would like answered by the time you finish this chapter.

3. How does the title of the Start-Up Activity relate to the subject of the chapter?

Section 1: What Is Climate? (p. 68)

4. When you step outside to see if there are rain clouds in the sky, you are checking the _____ (weather or climate)

5. If you look through your town's records for average temperature and rainfall over the past 10 years, you are looking at _____ data. (weather or climate)

REINFORCEMENT & VOCABULARY REVIEW

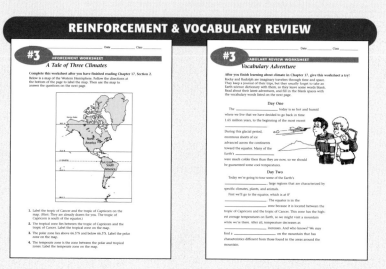

#3 REINFORCEMENT WORKSHEET
A Tale of Three Climates

Complete this worksheet after you have finished reading Chapter 17, Section 2. Below is a map of the Western Hemisphere. Follow the directions at the bottom of the page to label the map. Then use the map to answer the questions on the next page.

1. Label the tropic of Cancer and the tropic of Capricorn on the map. (Hint: They are already drawn for you. The tropic of Capricorn is south of the equator.)
2. The tropical zone lies between the tropic of Capricorn and the tropic of Cancer. Label the tropical zone on the map.
3. The polar zone lies above 66.5°N and below 66.5°S. Label the polar zone on the map.
4. The temperate zone is the zone between the polar and tropical zones. Label the temperate zone on the map.

#3 VOCABULARY REVIEW WORKSHEET
Vocabulary Adventure

After you finish learning about climate in Chapter 17, give this worksheet a try! Rocky and Rudolph are imaginary travelers through time and space. They keep a journal of their trips, but they usually forget to take an Earth science dictionary with them, so they leave some words blank. Read about their latest adventures, and fill in the blank spaces with the vocabulary words listed on the next page.

Day One
The _____ today is so hot and humid where we live that we have decided to go back in time 1.65 million years, to the beginning of the most recent

During this glacial period, enormous sheets of ice advanced across the continents toward the equator. Many of the Earth's _____ were much colder than they are now, so we should be guaranteed some cool temperatures.

Day Two
Today we're going to tour some of the Earth's _____, large regions that are characterized by specific climates, plants, and animals.
First we'll go to the equator, which is at 0° _____. The equator is in the _____ zone because it is located between the tropic of Capricorn and the tropic of Cancer. This zone has the highest average temperatures on Earth, so we might visit a mountain while we're there. After all, temperature decreases as _____ increases. And who knows? We may find a _____ on the mountain that has characteristics different from those found in the areas around the mountain.

SCIENCE PUZZLERS, TWISTERS & TEASERS

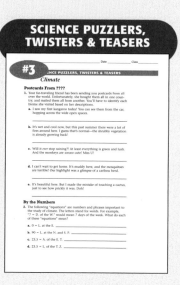

#3 SCIENCE PUZZLERS, TWISTERS & TEASERS
Climate

Postcards From ????
1. Your far-traveling friend has been sending you postcards from all over the world. Unfortunately, she bought them all in one country, and mailed them all from another. You'll have to identify each biome she visited based on her descriptions.

a. I saw my first kangaroo today! You can see them from the car, hopping across the wide open spaces.

b. It's wet and cool now, but this past summer there were a lot of fires around here. I guess that's normal—the shrubby vegetation is already growing back!

c. Will it ever stop raining?! At least everything is green and lush. And the monkeys are sooooo cute! Miss U!

d. I can't wait to get home. It's muddy here, and the mosquitoes are terrible! Our highlight was a glimpse of a caribou herd.

e. It's beautiful here. But I made the mistake of touching a cactus, just to see how prickly it was. Doh!

By the Numbers
2. The following "equations" are numbers and phrases important to the study of climate. The letters stand for words. For example, "7 = D. of the W." would mean 7 days of the week. What do each of these "equations" mean?

a. 0 = L. at the E. _____
b. 90 = L. at the N. and S. P. _____
c. 23.5 = A. of the E. T. _____
d. 23.5 = L. of the T. Z. _____

Review & Assessment

STUDY GUIDE

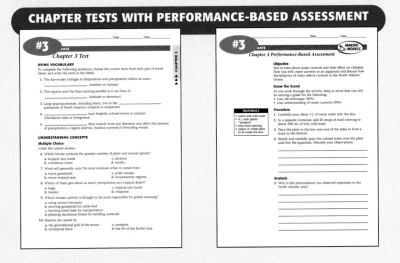

CHAPTER TESTS WITH PERFORMANCE-BASED ASSESSMENT

Lab Worksheets

WHIZ-BANG DEMONSTRATIONS

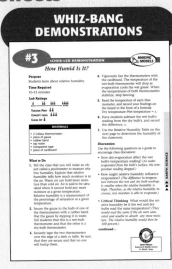

LONG-TERM PROJECTS & RESEARCH IDEAS

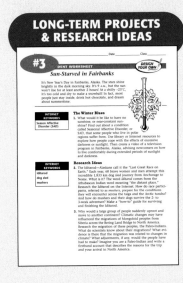

DATASHEETS FOR LABBOOK

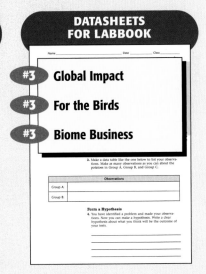

Applications & Extensions

CRITICAL THINKING & PROBLEM SOLVING

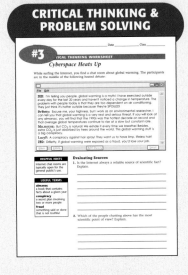

EYE ON THE ENVIRONMENT

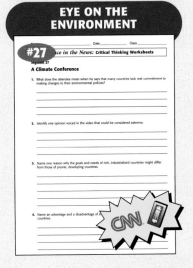

SCIENTISTS IN ACTION

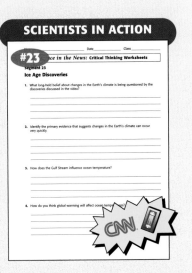

SECTION 1

What Is Climate?

▶ **Climatology**

The study of climate can be traced back to Greek scientists of the sixth century B.C. In fact, the word *climate* comes from the Greek word *klíma*, meaning "an inclination," such as the angle of the sun's rays. Climatology can be divided into three branches—global climatology, regional climatology, and physical climatology. Global climatology investigates the general circulation of wind and water currents around Earth. Regional climatology studies the characteristic weather patterns and related phenomena of a particular region. Physical climatology analyzes statistics concerning climatic factors such as temperature, moisture, wind, and air pressure.

▶ **Global Winds**

Global winds are patterns of air circulation that travel across the Earth. These winds include the trade winds, the prevailing westerlies, and the polar easterlies.

- In both hemispheres, the trade winds blow from 30° latitude to the equator. The Coriolis effect makes the trade winds curve to the right in the Northern Hemisphere, moving northeast to southwest. In the Southern Hemisphere, the trade winds curve to the left and move from southeast to northwest.

- The westerlies are found in both the Northern and Southern Hemispheres between 30° and 60° latitude. In the Northern Hemisphere, the westerlies blow from the southwest to the northeast. In the Southern Hemisphere, they blow from the northwest to the southeast.

- The polar easterlies extend from the poles to 60° latitude in both hemispheres. The polar easterlies blow from the northeast to the southwest in the Northern Hemisphere. In the Southern Hemisphere, these winds blow from the southeast to the northwest.

SECTION 2

Climates of the World

▶ **Climate Classification**

Because climate is a complicated and somewhat abstract concept, more than 100 classification models have been devised, which vary according to the data on which the classifications are based. For instance, there have been attempts to classify climates according to factors such as soil formation, rock weathering, and even effects on human comfort!

Polar
66.5°N — Temperate
23.5°N — Tropical
0° — Tropical
23.5°S — Temperate
66.5°S — Polar

- In 1966, Werner Terjung, an American geographer, developed a physiological climate classification. This system categorized climates according to their effects on people's comfort levels. The system focused on four factors that might affect human comfort—temperature, relative humidity, wind speed, and solar radiation.

The Köppen System

The most widely used climate classification system is the Köppen system. This system, named for Wladimir Köppen, the German botanist and climatologist who developed it, uses vegetation regions and average weather statistics to classify local climates. Each vegetation region is characterized by the natural vegetation that is predominant there. Critics have found fault with the Köppen system because it considers only average monthly temperatures and precipitation, ignoring other factors, such as winds, cloud cover, and daily temperature extremes.

SECTION 3

Changes in Climate

▶ Pangaea

In 1620, the British philosopher Francis Bacon noted that Africa and South America looked as if they could fit together like puzzle pieces. But it was not until the early twentieth century that the German meteorologist Alfred Wegener proposed a theory that all the continents were once one landmass. Wegener's hypothesis was supported by the existence of similar plant and animal fossils on different continents. Although his theory was initially ridiculed, Wegener was vindicated after World War II when sea-floor spreading and a mechanism for continental drift were discovered.

IS THAT A FACT!

- ☛ *Pangaea,* the name Wegener gave to the supercontinent, is Greek for "all earth."

▶ The Greenhouse Effect

Gases such as carbon dioxide, methane, and chlorofluorocarbons (CFCs) are known as greenhouse gases because they "trap" thermal energy in Earth's atmosphere by absorbing infrared radiation that would otherwise be emitted into space. Normal amounts of greenhouse gases, with the exception of CFCs (which are artificial chemicals), are necessary for life on Earth

because they keep Earth's average temperature at 15°C. Without them, Earth would be frozen; the average temperature would be about −18°C.

- ● CFCs are manufactured chemicals that contain chlorine, fluorine, and carbon. Before 1978, when the United States banned the use of CFCs, they were widely used as propellants in aerosol cans. They are also a component of plastic foam and of a coolant used in air conditioners. CFCs are extremely potent; one molecule of a CFC can absorb about 10,000 times more energy than a molecule of carbon dioxide.

- ● There is much less methane in the air than carbon dioxide. However, per molecule, methane absorbs 20 times more infrared radiation than carbon dioxide, and its concentration in the atmosphere is increasing. Methane is a natural product of animal digestion. A significant amount of the methane released into the atmosphere is produced by livestock. Methane is also produced as microorganisms decompose organic material.

IS THAT A FACT!

- ☛ Burning 1 gal of gasoline produces 9 kg of carbon dioxide.

- ☛ Using 1 kWh of electrical energy from a coal-fired power plant produces 5.5 kg of carbon dioxide.

For background information about teaching strategies and issues, refer to the *Professional Reference for Teachers.*

CHAPTER 3

Climate

 Pre-Reading Questions

Students may not know the answers to these questions before reading the chapter, so accept any reasonable response.

Suggested Answers

1. Weather is the condition of the atmosphere at a certain time or place. Climate is the average weather conditions over a long period of time.

2. Answers will vary. The increased use of fossil fuels may cause the climate to become warmer. This is because the carbon dioxide released into the atmosphere traps heat.

CHAPTER 3

Climate

Sections

 **Pre-Reading Questions**

1. What is the difference between weather and climate?
2. List ways in which human influences such as pollution and technology can affect climate.

66

A HOT NEW HOME

Snow macaques normally live in cold pine forests in the mountains of Japan. However, in 1972, a group of these monkeys was relocated to a ranch in southern Texas. The monkeys were forced to adapt to a radically different climate and environment, which meant learning how to live with higher temperatures, different plants, and different animals. In this chapter, you will learn about the factors that affect climate and about the different environments found in each climate.

internet connect

 HRW On-line Resources

go.hrw.com
For worksheets and other teaching aids, visit the HRW Web site and type in the keyword: **HSTCLM**

 sciLINKS **NSTA**

www.scilinks.com
Use the *sci*LINKS numbers at the end of each chapter for additional resources on the **NSTA** Web site.

 Smithsonian Institution

www.si.edu/hrw
Visit the Smithsonian Institution Web site for related on-line resources.

 CNN fyi.com

www.cnnfyi.com
Visit the CNN Web site for current events coverage and classroom resources.

WHAT'S YOUR ANGLE?

Because the Earth is round, the sun's solar rays strike the Earth's surface at different angles. Try this activity to find out how the amount of solar energy received at the equator differs from the amount received at the poles.

Procedure

1. Plug in a **lamp,** and position it 30 cm from a **globe.**

2. Point the lamp so that the light shines directly on the globe's equator.

3. Using **adhesive putty,** attach a **thermometer** to the globe's equator in a vertical position. Attach **another thermometer** to the globe's north pole so that the tip points toward the lamp.

4. Record the temperature reading of each thermometer in your ScienceLog.

5. Turn on the lamp, and let the light shine on the globe for 3 minutes.

6. When the time is up, turn off the lamp, and record the temperature reading of each thermometer again.

Analysis

7. Was there a difference between the final temperature at the globe's north pole and that at the globe's equator? If so, what was it?

67

WHAT'S YOUR ANGLE?

MATERIALS

FOR EACH GROUP:
• lamp
• globe
• adhesive putty
• 2 thermometers

Safety Caution

Remind students to review all safety cautions and icons before beginning this lab activity. Students should not touch the lamp's bulb while it is on or immediately after it has been turned off.

Teacher's Notes

If you have time, encourage students to repeat the experiment, positioning one thermometer at the equator and one at the South Pole. Have them compare their results. Students could also attempt to simulate how Earth's orbit affects the seasons.

Answer to START-UP Activity

7. The final temperature at the globe's North Pole was cooler than the final temperature at the globe's equator. This is because the globe's equator received more direct energy from the light bulb than the globe's North Pole received.

Focus

What Is Climate?

In this section, students learn the difference between weather and climate. They examine how latitude, prevailing winds, geography, and ocean currents affect an area's climate.

Bellringer

Have students imagine they have entered a contest for a free trip to a place with a perfect climate. To win, they need to describe in 25 words or less their idea of a perfect climate.

Sheltered English

1 Motivate

DISCUSSION

Tell students that in the early 1900s, a geographer named Ellsworth Huntington conducted controversial research to see if he could determine the ideal climate for human beings—the type of climate that would result in optimal physical and mental well-being. He concluded that a climate with considerable daily and seasonal weather changes and an average temperature of 18°C was ideal. Ask students whether they agree or disagree with Huntington and why.

Directed Reading Worksheet Section 1

Terms to Learn

weather	prevailing winds
climate	elevation
latitude	surface currents

What You'll Do

◆ Explain the difference between weather and climate.
◆ Identify the factors that determine climates.

What Is Climate?

You have just received a call from a friend who is coming to visit you tomorrow. He is wondering what clothing to bring and wants to know about the current weather in your area. You step outside, check to see if there are rain clouds in the sky, and note the temperature. But what if your friend asked you about the climate in your area? What is the difference between weather and climate?

The main difference between weather and climate has to do with time. **Weather** is the condition of the atmosphere at a particular time and place. Weather conditions vary from day to day. **Climate,** on the other hand, is the average weather conditions in an area over a long period of time. Climate is determined by two main factors, temperature and precipitation. Study the map in **Figure 1,** and see if you can describe the climate in northern Africa.

Figure 1 *How does the climate in northern Africa differ from the climate where you live?*

Multicultural CONNECTION

Weather and climate have inspired a great number of rhymes, greetings, sayings, and other folklore. For example, in the hot, wet climate of Venezuela, indigenous people sometimes greet each other by saying, "How have the mosquitoes used you?"

Russia's cold climate inspired the saying, "There's no bad weather, only bad clothing." Invite students to interview friends and relatives or research weather and climate folklore in another country. Have them share their findings with the class.

As you can see in **Figure 2,** if you were to take a trip around the world, or even across the United States, you would experience different climates. For example, if you visited the Texas coast in the summer, you would find it hot and humid. But if you visited interior Alaska during the summer, it would probably be much cooler and less humid. Why are the climates so different? The answer is complicated. It includes factors such as latitude, wind patterns, geography, and ocean currents.

Figure 2 *Summer in Texas is different from summer in Alaska.*

Latitude

Think of the last time you looked at a globe. Do you recall the thin horizontal lines that circle the globe? These horizontal lines are called lines of latitude. **Latitude** is the distance north or south, measured in degrees, from the equator. In general, the temperature of an area depends on its latitude. The higher the latitude is, the colder the climate is. For example, one of the coldest places on Earth, the North Pole, is at 90° north of the equator. On the other hand, the equator, which has a latitude of 0°, is hot.

It's Hot! It's Not! Why are there such temperature differences at different latitudes? The answer has to do with solar energy. Solar energy heats the Earth. Latitude determines the amount of solar energy a particular area receives. You can see how this works in **Figure 3.** Notice that the sun's rays hit the area around the equator directly, at nearly a 90° angle. At this angle, a small area of the Earth's surface receives more direct solar energy, resulting in high temperatures. Near the poles, however, the sun's rays strike the surface at a lesser angle than at the equator. This lesser angle spreads the same amount of solar energy over a larger area, resulting in lower temperatures.

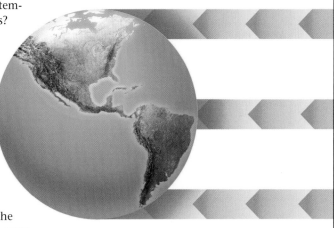

Figure 3 *The sun's rays strike the Earth's surface at different angles because the surface is curved.*

69

internet**connect**

*SCI*LINKS

NSTA

TOPIC: What Is Climate?
GO TO: www.scilinks.org
*sci***LINKS NUMBER:** HSTE405

MATH and MORE

Each degree or line of latitude is approximately 111 km, and there are 180 lines of latitude circling the Earth. Have students calculate the circumference of the Earth at the poles. (111 km/line × 180 = 19,980 km) Then multiply by 2, because the Earth is a sphere. (39,960 km)

CONNECT TO GEOGRAPHY

Monsoons are recurrent global weather patterns that dramatically affect the populations, economies, and environment of South Asia. The wet summer monsoon usually begins mid-June, when temperatures rise sharply in Asia's interior, causing the air above the land to warm and rise. This creates a low pressure area that draws warm, moist air inland from the Indian and Pacific oceans. This moisture-laden air cools as it moves across the continent, and heavy rains, thunderstorms, and flooding occur. The heaviest rains occur where this air mass meets the foothills of the Himalayas. During a winter monsoon, the interior of Asia cools rapidly. This cool, dense air creates an immense high pressure center, forcing cool, dry air to flow outward toward the oceans. As the air mass travels, it warms and becomes even drier. This results in warm, dry winters. Encourage students to write a report about the effect of monsoons on the environment, economy, and people of South Asia.

COOPERATIVE LEARNING

Ask two volunteers to act as the Earth and the sun. Give the "sun" a flashlight and the "Earth" a globe with a half-meridian mounting. Turn off the lights, and have the volunteers sit on the floor. Ask the "sun" to shine the flashlight on the globe, and ask the "Earth" to slowly spin the globe. Have the class notice which parts of the globe are most exposed to the light. Next, have the "Earth" make a complete revolution slowly around the "sun" while at the same time rotating the globe. Make sure that the volunteer always keeps the axis of the globe oriented in the same direction. Stop the "Earth" at each season so students can observe the flashlight's rays on the two hemispheres. If necessary, repeat this activity with other volunteers.

Sheltered English

MISCONCEPTION ///ALERT\\\

Students may think that the distance between the Earth and the sun determines the seasons. Actually, the Earth is closer to the sun during the Northern Hemisphere's winter. Remind them that the Earth's tilt as it orbits the sun determines where solar radiation is concentrated and thus determines the seasons.

Answer to Self-Check

Australia has summer during our winter months, December–February.

Teaching Transparency 175 "Seasons, Latitude, and the Tilt of the Earth"

BRAIN FOOD

The polar regions receive almost 24 hours of daylight each day in the summer and almost 24 hours of darkness each day in the winter.

Seasons and Latitude In most places in the United States, the year consists of four seasons. Winter is probably cooler than summer where you live. But there are places in the world that do not have such seasonal changes. For example, areas near the equator have approximately the same temperatures and same amount of daylight year-round. **Figure 4** shows how latitude determines the seasons.

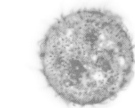
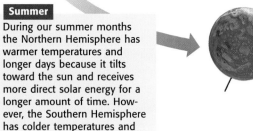

March 21

June 21

December 21

September 22

Winter
During our winter months the Southern Hemisphere has higher temperatures and longer days because it tilts toward the sun and receives more direct solar energy. The Northern Hemisphere has lower temperatures and shorter days because it tilts away from the sun.

Summer
During our summer months the Northern Hemisphere has warmer temperatures and longer days because it tilts toward the sun and receives more direct solar energy for a longer amount of time. However, the Southern Hemisphere has colder temperatures and shorter days because it is tilted away from the sun.

Figure 4 The Earth is tilted on its axis at a 23.5° angle. This tilt affects how much solar energy an area receives as the Earth moves around the sun.

✓ Self-Check

During what months does Australia have summer? (See page 136 to check your answer.)

IS THAT A FACT!

Tutunendo, Colombia, is the rainiest place in the world. It averages almost 12 m of rain per year. Cherrapunji, India, received more than 9 m of rain in 1 month—the most rainfall ever recorded in a month. The hottest day ever recorded occurred in Libya, where it reached 58°C (136°F) in 1922. Antarctica holds the record as the coldest place on Earth, with temperatures reaching –89°C (–128°F).

Prevailing Winds

Prevailing winds are winds that blow mainly from one direction. These winds influence an area's moisture and temperature. Before you learn how the prevailing winds affect climate, take a look at **Figure 5** to learn about some of the basic properties of air.

Figure 5 *Because warm air is less dense, it tends to rise. Cooler, denser air tends to sink.*

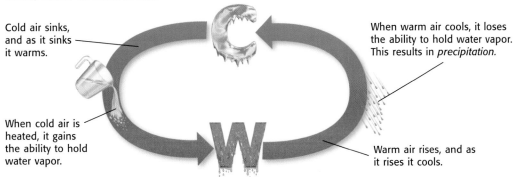

Cold air sinks, and as it sinks it warms.

When warm air cools, it loses the ability to hold water vapor. This results in *precipitation.*

When cold air is heated, it gains the ability to hold water vapor.

Warm air rises, and as it rises it cools.

Prevailing winds affect the amount of precipitation that a region receives. If the prevailing winds form from warm air, they will carry moisture. If the prevailing winds form from cold air, they will probably be dry.

The amount of moisture in prevailing winds is also affected by whether the winds blow across land or across a large body of water. Winds that travel across large bodies of water absorb moisture. Winds that travel across land tend to be dry. Even if a region borders the ocean, the area might be dry if the prevailing winds blow across the land, as shown in **Figure 6.**

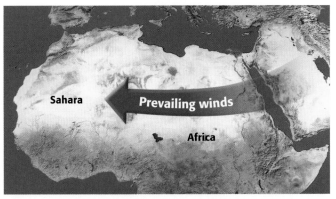

Sahara

Prevailing winds

Africa

Figure 6 *The Sahara Desert, in northern Africa, is extremely dry because of the dry prevailing winds that blow across the continent.*

QuickLab

A Cool Breeze

1. Hold a **thermometer** next to the top edge of a **cup of water** containing **two ice cubes.** Read the temperature next to the cup.

2. Have your lab partner fan the surface of the cup with a **paper fan.** Read the temperature again. Has the temperature changed? Why? Record your answer in your ScienceLog.

TRY at HOME

Multicultural CONNECTION

Charles Edward Anderson was the first African American to receive a Ph.D. in meteorology. Anderson began his career in meteorology during World War II. He was a captain in the Air Force and served as a weather officer for the Tuskegee Airmen Regiment. He earned his Ph.D. in 1960 from the Massachusetts Institute of Technology. His work focused on cloud physics, the forecasting of severe storms, and weather on other planets.

Activity

Using a physical map, locate the mountain ranges in the United States. Does climate vary from one side of a mountain range to the other? If so, what does this tell you about the climatic conditions on either side of the mountain? From what direction are the prevailing winds blowing?

TRY at HOME

Geography

Mountains can influence an area's climate by affecting both temperature and precipitation. For example, Kilimanjaro, the tallest mountain in Africa, has snow-covered peaks year-round, even though it is only about 3° (320 km) south of the equator. Temperatures on Kilimanjaro and in other mountainous areas are affected by elevation. **Elevation** is the height of surface landforms above sea level. As the elevation increases, the atmosphere becomes less dense. When the atmosphere is less dense, its ability to absorb and hold thermal energy is reduced and temperatures are therefore lower.

Mountains also affect the climate of nearby areas by influencing the distribution of precipitation. **Figure 7** shows how the climates on two sides of a mountain can be very different.

Figure 7 *Mountains block the prevailing winds from blowing across a continent, changing the amount of moisture the wind carries.*

The Wet Side
Mountains force air to rise. The air cools as it rises, releasing moisture as snow or rain. The land on the windward side of the mountain is usually green and lush due to the wind losing its moisture.

The Dry Side
After dry air crosses the mountain, the air begins to sink, warming and absorbing moisture as it sinks. The dry conditions created by the sinking, warm air usually produce a desert. This side of the mountain is in a *rain shadow.*

72

SCIENCE HUMOR

Western and eastern Oregon have very different climates because the Cascade Mountains divide the state. Oregonians living east of the Cascades complain, "It's so dry, the jackrabbits carry canteens." West of the Cascades, people say, "It's so wet, folks don't tan, they rust!"

IS THAT A FACT!

Large lakes, such as the Great Lakes, in the United States and Canada, and Lake Victoria, in east-central Africa, affect local climates. This phenomenon, called *the lake effect,* helps keep the surrounding land cooler in the summer and warmer in the winter.

Ocean Currents

Because of water's ability to absorb and release thermal energy, the circulation of ocean surface currents has an enormous effect on an area's climate. **Surface currents,** which can be either warm or cold, are streamlike movements of water that occur at or near the surface of the ocean. **Figure 8** shows the pattern of the major warm and cold ocean surface currents.

Current Events As surface currents move, they carry warm or cool water to different locations. The surface temperature of the water affects the temperature of the air above it. Warm currents heat the surrounding air and cause warmer temperatures, while cool currents cool the surrounding air and cause cooler temperatures. For example, the Gulf Stream current carries warm water northward off the east coast of North America past Iceland, an island country located just below the Arctic Circle. The warm water from the Gulf Stream heats the surrounding air, creating warmer temperatures in southern Iceland. Iceland experiences milder temperatures than Greenland, its neighboring country, where the climate is not influenced by the Gulf Stream.

Science CONNECTION

What is El Niño? Can it affect our health? Turn to page 94 to find out.

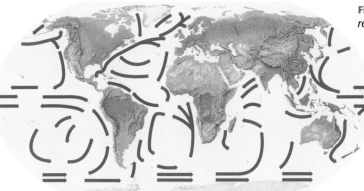

Figure 8 *The red arrows represent the movement of warm surface currents. The blue arrows represent the movement of cold surface currents.*

SECTION REVIEW

1. What is the difference between weather and climate?

2. How do mountains affect climate?

3. Describe how air temperature is affected by ocean surface currents.

4. **Analyzing Relationships** How would seasons be different if the Earth did not tilt on its axis?

internetconnect

SCiLINKS
NSTA

TOPIC: What Is Climate?
GO TO: www.scilinks.org
*sci*LINKS NUMBER: HSTE405

73

Focus

Climates of the World

In this section, students learn the location and the characteristics of the three major climate zones and the different types of biomes that are found in each zone.

Bellringer

Ask students to describe in their ScienceLog differences between the plant life where they live and the plant life in an area they have visited. Ask students to think about how climate influences the vegetation in these areas. Sheltered English

1 Motivate

ACTIVITY

Writing Have each student choose a country to focus on for this section. Students should record the area's latitude and geographic characteristics in their ScienceLog. Tell students to imagine they are on a fact-finding mission concerning the area's climate, the people that live there, and the plant and animal life of the region. Have them record evidence about the area's climate as they read the section.

Teaching Transparency 178 "Climate Zones of the Earth" "The Earth's Land Biomes"

Directed Reading Worksheet Section 2

SECTION **2**
READING WARM-UP

Terms to Learn

biome	evergreens
tropical zone	polar zone
temperate zone	microclimate
deciduous	

What You'll Do

◆ Locate and describe the three major climate zones.
◆ Describe the different biomes found in each climate zone.

Climates of the World

Have you ever wondered why the types of plants and animals in one part of the world are different from those found in another part? One reason involves climate. Plants and animals that have adapted to one climate may not be able to live in another climate. For instance, frogs do not live in Antarctica.

The three major climate zones of Earth—tropical, temperate, and polar—are illustrated in **Figure 9**. Each zone has a temperature range that relates to its latitude. However, in each of these zones there are several types of climates due to differences in the geography and the amount of precipitation. Because of the various climates in each zone, there are different biomes. A **biome** is a large region characterized by a specific type of climate and the plants and animals that live there.

Figure 10 shows the distribution of the Earth's land biomes. In this section we will review each of the three major climate zones and the biomes that are found in each zone.

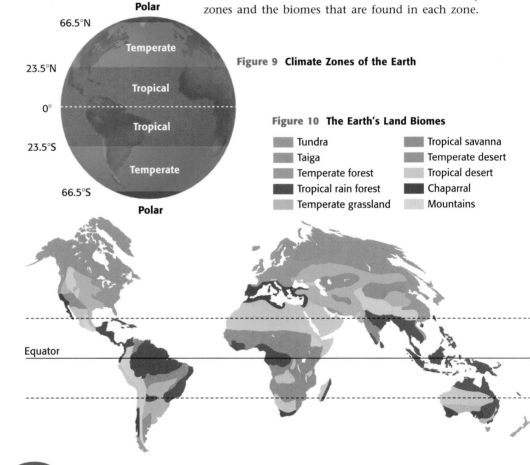

Figure 9 **Climate Zones of the Earth**

Figure 10 **The Earth's Land Biomes**

Tundra	Tropical savanna
Taiga	Temperate desert
Temperate forest	Tropical desert
Tropical rain forest	Chaparral
Temperate grassland	Mountains

WEIRD SCIENCE

In addition to having land biomes, Earth also has marine biomes. It is impossible to distinguish biomes in the oceans by latitude, however. Marine biomes are determined by water depth. Some of the animals that inhabit the deeper biomes have very interesting adaptations. For example, the anglerfish, which lives in total darkness, has a clever hunting strategy: a luminescent "lure" trails from the fish's jaw and attracts prey within reach of its enormous, sharp teeth.

The Tropical Zone

The **tropical zone,** or the *Tropics,* is the warm zone located around the equator, as shown in **Figure 11.** This zone extends from the tropic of Cancer to the tropic of Capricorn. As you have learned, latitudes in this zone receive the most solar radiation. Temperatures are therefore usually hot, except at high elevations. Within the tropical zone there are three types of biomes—tropical rain forest, tropical desert, and tropical savanna. **Figure 12** shows the distribution of these biomes.

Figure 11   The Earth's Tropical Zone

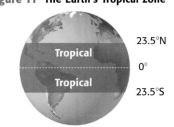

23.5°N

0°

23.5°S

Figure 12   Biomes of the Tropical Zone

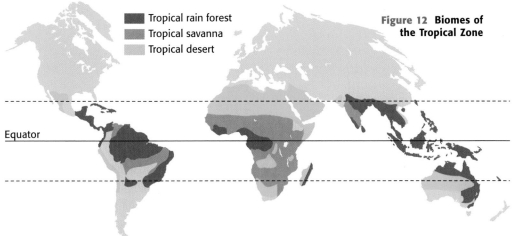

- Tropical rain forest
- Tropical savanna
- Tropical desert

Equator

Tropical Rain Forest   Tropical rain forests are always warm and wet. Because they are located near the equator, they receive strong sunlight year-round, causing little difference between seasons.

Tropical rain forests contain the greatest number of plant and animal species of any biome. But in spite of the lush vegetation, shown in **Figure 13,** the soil in rain forests is poor. The rapid decay of plants and animals returns nutrients to the soil, but these nutrients are quickly absorbed and used by the plants. The nutrients that are not immediately used by the plants are washed away by the heavy rains, leaving soil that is thin and nutrient poor.

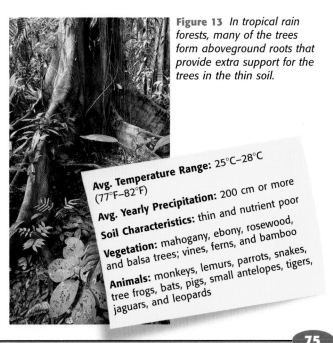

Figure 13 *In tropical rain forests, many of the trees form aboveground roots that provide extra support for the trees in the thin soil.*

Avg. Temperature Range: 25°C–28°C (77°F–82°F)
Avg. Yearly Precipitation: 200 cm or more
Soil Characteristics: thin and nutrient poor
Vegetation: mahogany, ebony, rosewood, and balsa trees; vines, ferns, and bamboo
Animals: monkeys, lemurs, parrots, snakes, tree frogs, bats, pigs, small antelopes, tigers, jaguars, and leopards

It's not a jungle out there. The popular image of a tropical rain forest is of dense jungle undergrowth, but this occurs only along rivers or places that humans have cleared. Students may also confuse rain forests with forests in monsoon-climate regions. While rain forests have a fairly steady rate of precipitation, monsoon forests have a rainy season and a dry season. Rainfall during the rainy season may be measured in meters. During the dry season, the monsoon forest may receive little or no rainfall for many months. In fact, many monsoon-forest plants have some of the same adaptations for dry conditions that desert plants have.

REAL-WORLD CONNECTION

Deserts are expanding at an accelerating rate. In the last 100 years, the estimated area of land occupied by deserts rose from 9.4 percent to 23.3 percent. Many factors have contributed to this, including climatic shifts, overgrazing, and overuse of the land through inefficient agricultural practices. As a class, find out what is being done to stop desertification in Western Africa and other areas of the world.

COOPERATIVE LEARNING

Many people believe that conditions in the desert biomes are so extreme that they are nearly devoid of life. In fact, many different kinds of deserts can support thriving ecological communities. Have students choose a tropical or temperate desert and create a poster display to teach the class about its location, the organisms that inhabit the desert, and other facts about the desert. Students could study the Gobi Desert, the Sahara Desert, the Great Sandy Desert, the Sonoran Desert, the Patagonian Desert, the Namib Desert, or another desert of their choice.

Homework

Research Draw students' attention to an important distinction between the vegetation in Old World and New World tropical deserts. Cactus is a succulent found in the New World. Euphorbia is a succulent found in the Old World. Have students find photographs of these types of plants and share them with the class.

Most desert rodents, such as the kangaroo rat, hide in burrows during the day and are active at night, when the temperatures are cooler.

Biology
CONNECTION

Some desert animals, such as the spadefoot toad, survive the scorching summer heat by burying themselves in the ground and sleeping through the dry season.

Tropical Deserts A desert is an area that receives less than 25 cm of rainfall per year. Because of this low yearly rainfall, deserts are the driest places on Earth. Desert plants, shown in **Figure 14,** are adapted to survive in a place with little water.

Deserts can be divided into hot deserts and cold deserts. The majority of hot deserts, such as the Sahara, in Africa, are tropical deserts. Hot deserts are caused by cool sinking air masses. Daily temperatures in tropical deserts vary from very hot daytime temperatures (50°C) to cool nighttime temperatures (20°C). Winters in hot deserts are usually mild. Because of the dryness, the soil is poor in organic matter, which fertilizes the soil. The dryness makes it hard to break down dead organic matter.

Avg. Temperature Range: 16°C–50°C (61°F–120°F)

Avg. Yearly Precipitation: 0–25 cm

Soil Characteristics: poor in organic matter

Vegetation: succulents (cactus and euphorbia), shrubs, thorny trees

Animals: kangaroo rats, lizards, scorpions, snakes, birds, bats, toads

Figure 14 *Plants called succulents have adapted to dry conditions by developing fleshy stems and leaves to store water and a waxy coating to prevent water loss. A cactus is a type of succulent.*

✓ Self-Check

If desert soil is so nutrient rich, why are deserts not suitable for agriculture? *(See page 136 to check your answer.)*

Answer to Self-Check

Because of its dryness, desert soil is poor in organic matter, which fertilizes the soil. Without this natural fertilizer, crops would not be able to grow.

IS THAT A FACT!

The world's largest desert, the Sahara, covers more than 9 million square kilometers—about the size of the United States. In contrast, the largest desert in the United States, is the Mojave Desert. It covers 38,900 km², which is nearly twice the size of New Jersey.

Tropical Savannas Tropical savannas, sometimes referred to as grasslands, are dominated by tall grasses, with trees scattered here and there. **Figure 15** is a photo of an African savanna. The climate is usually very warm, with a dry season that lasts four to eight months followed by short periods of rain. Savanna soils are generally nutrient poor, but grass fires, which are common during the dry season, leave the soils nutrient enriched.

Many plants have adapted to fire and use it to reproduce. Grasses sprout from their roots after the upper part of the plant is burned. The seeds of some plant species require fire in order to grow. For example, some species need fire to break open the seed's outer skin. Only after this skin is broken can the seed grow. Other species drop their seeds at the end of fire season. The heat from the fire triggers the plants to drop their seeds into the newly enriched soil.

Avg. Temperature Range: 27°C–32°C (80°F–90°F)

Avg. Yearly Precipitation: 100 cm

Soil Characteristics: generally nutrient poor

Vegetation: tall grasses (3–5 m), trees, thorny shrubs

Animals: gazelles, rhinoceroses, giraffes, lions, hyenas, ostriches, crocodiles, elephants

Figure 15 *The grass of a tropical savanna is 3–5 m tall, much taller than that of a temperate grassland.*

SECTION REVIEW

1. What are the soil characteristics of a tropical rain forest?

2. In what way has savanna vegetation adapted to fire?

3. **Summarizing Data** How do each of the tropical biomes differ?

internet**connect**

*SCi*LINKS
NSTA

TOPIC: Climates of the World
GO TO: www.scilinks.org
*sci*LINKS NUMBER: HSTE410

77

▼ **Answers to Section Review**

1. The soil in a tropical rain forest is thin and nutrient poor. Nutrients are rapidly returned to the soil, but these nutrients are quickly absorbed and used by the plants. The remaining nutrients are washed away by heavy rains.

2. Many plants require fire to reproduce. The seeds of some plants require fire to break open the seed's outer skin so the plant can grow. The heat from the fire triggers other plants to drop their seeds into the newly enriched soil.

3. Answers will vary. Accept all reasonable responses. The tropical biomes differ in the amount of precipitation they receive and in their average temperature. This in turn affects the vegetation, the soil type, and the animals that live in each biome.

READING STRATEGY

Prediction Guide If your community is in the continental United States, with the exception of Alaska, point out to students that they live in the temperate climate zone. The temperate climate zone is made up of four biomes: temperate forest, temperate grassland, chaparral, and temperate desert.

Write the following sentence on the board and have students copy it into their ScienceLog:

"I think we live in the _____ biome because _____."

Have them complete the sentence with the name of the biome and some ideas that support their statement.

MISCONCEPTION ///ALERT\\\

You don't need to travel to the tropics to find a rain forest. Western Washington state is home to the largest temperate rain forest in the world. Moss-covered trees more than 500 years old stand 60 m tall and are 5 m in diameter. The ground is covered by ferns, moss, salmonberries, and the thorny Hercules's club. The growth is not as diverse as the tropical forests, but it is every bit as lush. The forest receives 380 cm of rain a year! Have students find out more about this remarkable ecosystem and the efforts to preserve it.

internet connect

SCLINKS **TOPIC:** Climates of the World
GO TO: www.scilinks.org
NSTA *sci*LINKS NUMBER: HSTE410

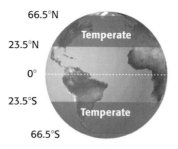

66.5°N
23.5°N
0°
23.5°S
66.5°S

Figure 16 The Earth's Temperate Zones

The Temperate Zone

The **temperate zone,** as shown in **Figure 16,** is the climate zone between the Tropics and the polar zone. Temperatures in the temperate zone tend to be moderate. The continental United States is in the temperate zone, which includes the following four biomes: temperate forest, temperate grassland, chaparral, and temperate desert. **Figure 17** shows the distribution of the biomes found in the temperate zone.

Figure 17 Biomes of the Temperate Zone

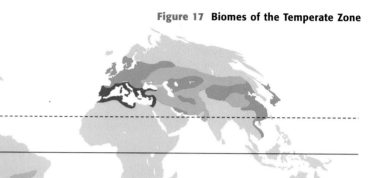

Equator

■ Temperate forest
■ Temperate grassland
■ Temperate desert
■ Chaparral

Figure 18 *Deciduous trees have leaves that change color and drop when temperatures become cold.*

Avg. Temperature Range: 0°C–28°C (32°F–82°F)

Avg. Yearly Precipitation: 76–250 cm

Soil Characteristics: very fertile, organically rich

Vegetation: deciduous and evergreen trees, shrubs, herbs

Animals: deer, bears, boars, badgers, squirrels, wolves, wild cats, red foxes, owls, and many other birds

Temperate Forests The temperate forest biomes tend to have very high amounts of rainfall and seasonal temperature differences. Because of these distinct seasonal changes, summers are usually warm and winters are usually cold. The largest temperate forests are deciduous, such as the one shown in **Figure 18.** **Deciduous** trees are trees that lose their leaves when the weather becomes cold. These trees tend to be broad-leaved. The soils in deciduous forests are usually quite fertile because of the high organic content contributed by decaying leaves that drop every winter.

Another type of temperate forest is the evergreen forest. **Evergreens** are trees that keep their leaves year-round. Evergreens can be either broad-leaved trees or needle-leaved trees, such as pine trees. Mixed forests of broad-leaved and needle-leaved trees can be found in humid climates, such as Florida, where winter temperatures rarely fall below freezing.

78

CONNECT TO ENVIRONMENTAL SCIENCE

The settlement of humans in temperate forests around the world has greatly fragmented these areas. Much of the temperate forest has been converted to agricultural land or logged. Fragmentation has a negative impact on many plants and animals that have certain habitat requirements. In fact, fragmentation has led to the extinction of many species. The temperate forest is also vulnerable to air pollution resulting from industrial activity. Acid precipitation and ozone has damaged entire forests, either killing the trees or making them more susceptible to disease. Have students write a persuasive essay explaining the importance of conserving Earth's temperate forests.

Temperate Grasslands Temperate grasslands, such as those shown in **Figure 19,** occur in regions that receive too little rainfall for trees to grow. This biome has warm summers and cold winters. The temperate grasslands are known by many local names—the *prairies* of North America, the *steppes* of Eurasia, the *veldt* of Africa, and the *pampas* of South America. Grasses are the most common type of vegetation found in this biome. Because grasslands have the most fertile soils of all biomes, much of the temperate grassland has been plowed to make room for croplands.

Avg. Temperature Range: –6°C–26°C (21°F–78°F)

Avg. Yearly Precipitation: 38–76 cm

Soil Characteristics: most fertile soils of all biomes

Vegetation: grasses

Animals: large grazing animals, including the bison of North America, the kangaroo of Australia, and the antelope of Africa

Figure 19 *The world's grasslands once covered about 42 percent of Earth's total land surface. Today they occupy only about 12 percent of the Earth's surface.*

Chaparrals Chaparral regions, as shown in **Figure 20,** have cool, wet winters and hot, dry summers. The vegetation is mainly evergreen shrubs, which are short, woody plants with thick, waxy leaves. The waxy leaves are adaptations that help prevent water loss in dry conditions. These shrubs grow in rocky, nutrient-poor soil. Like tropical-savanna vegetation, chaparral vegetation has adapted to fire. In fact, some plants, such as chamise, can grow back from their roots after a fire.

Avg. Temperature Range: 11°C–26°C (51°F–78°F)

Avg. Yearly Precipitation: 48–56 cm

Soil Characteristics: rocky, nutrient-poor soils

Vegetation: evergreen shrubs, scrubby trees, herbs

Animals: ground squirrels, deer, elk, mountain lions, coyotes, wolves

Figure 20 *Some plant species found in chaparral produce substances that help them catch on fire. These species require fire to reproduce.*

79

IS THAT A FACT!

Some grasses have defensive adaptations against grazing animals. Cordgrass, also known as rip gut, is found in marshy grassland areas and has sharp, hooklike barbs on its leaves that can easily cut an animal's mouth or a person's hands.

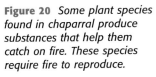
Science Skills Worksheet
"Finding Useful Sources"

BRAIN FOOD

Chile's arid northern desert land is one of the driest places on Earth. It receives so little rainfall that the yearly average is listed as "immeasurable." Surprisingly, people live there. They get drinking water by harvesting the fog. The village of Chungungo has built 75 fog-catching nets that supply 11,000 L of clean water a day. The nets, which look like giant volleyball nets, are positioned in the hills above the town. As the mountain fog passes through the nets, beads of water collect and are channeled to a large pipeline that supplies the village with water. Scientists believe this technology could be used in 30 other countries to supply safe and inexpensive water for drinking and agriculture.

Homework

Graphing Have students construct a bar graph that compares the average yearly precipitation ranges for the nine biomes discussed in the text. Have them use their graph to determine which biomes receive the most rain, which biomes receive the least rain, and which biome has the widest variation in annual precipitation. Suggest that students obtain yearly precipitation records for your region and compare them with the information in their graph.

Avg. Temperature Range: 1°C–50°C (34°F–120°F)

Avg. Yearly Precipitation: 0–25 cm

Soil Characteristics: poor in organic matter

Vegetation: succulents (cactus), shrubs, thorny trees

Animals: kangaroo rats, lizards, scorpions, snakes, birds, bats, toads

Temperate Deserts The temperate desert biomes, like the one shown in **Figure 21,** tend to be cold deserts. Like all deserts, cold deserts receive less than 25 cm of rainfall annually. Temperate deserts can be very hot in the daytime, but—unlike hot deserts—they tend to be very cold at night.

Figure 21 *The Great Basin Desert is in the rain shadow of the Sierra Nevada.*

The temperatures sometimes drop below freezing. This large change in temperature between day and night is caused by low humidity and cloudless skies. These conditions allow for a large amount of energy to reach, and thus heat, the Earth's surface during the day. However, these same characteristics allow the energy to escape at night, causing temperatures to drop. You probably rarely think of snow and deserts together, but temperate deserts often receive light snow during the winter.

Temperate deserts are dry because they are generally located inland, far away from a moisture source, or are located on the rain-shadow side of a mountain range.

The Polar Zone

The **polar zone** includes the northernmost and southernmost climate zones, as shown in **Figure 22.** Polar climates have the coldest average temperatures. The temperatures in the winter stay below freezing, and the temperatures during the summer months remain chilly. **Figure 23,** on the next page, shows the distribution of the biomes found in the polar zone.

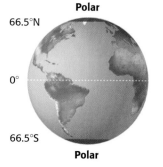

Polar
66.5°N

0°

66.5°S
Polar

Figure 22 The Earth's Polar Zones

WEIRD SCIENCE

Lichens are primitive organisms that thrive in the polar zone. Some lichens in the Arctic have been determined to be 4,500 years old. To protect themselves from the cold, some lichens live 2 cm inside rocks! Despite their ability to survive in extremely harsh arctic conditions, most lichens have an extremely low tolerance for sulfur dioxide air pollution. As a result, they are usually not found in industrialized areas.

Figure 23 Biomes of the Polar Zone

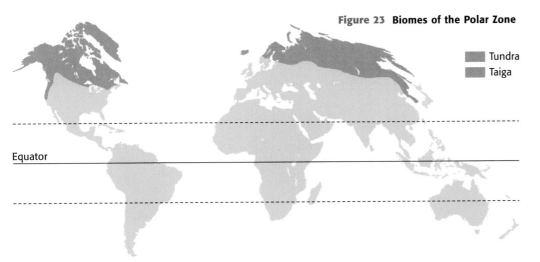

Equator

Tundra Next to deserts, the tundra, as shown in **Figure 24,** is the driest place on Earth. This biome has long, cold winters with almost 24 hours of night and short, cool summers with almost 24 hours of daylight. In the summer, only the top meter of soil thaws. Underneath the thawed soil lies a permanently frozen layer of soil, called *permafrost.* This frozen layer prevents the water in the thawed soil from draining. Because of the poor drainage, the upper soil layer is muddy and is therefore an excellent breeding ground for insects, such as mosquitoes. Many birds migrate to the tundra during the summer to feed on the insects.

Environment
CONNECTION

Subfreezing climates contain almost no decomposing bacteria. The well-preserved body of John Torrington, a member of an expedition that explored the Northwest Passage in Canada in the 1840s, was uncovered in 1984, appearing much as it did when he died, more than 140 years earlier.

Avg. Temperature Range: −27°C–5°C (−17°F–41°F)

Avg. Yearly Precipitation: 0–25 cm

Soil Characteristics: frozen

Vegetation: mosses, lichens, sedges, and dwarf trees

Animals: rabbits, lemmings, reindeer, caribou, musk oxen, wolves, foxes, birds, and polar bears

Figure 24 *In the tundra, mosses and lichens cover rocks. Dwarf trees grow close to the ground to protect themselves from strong winds and to absorb energy from the Earth's sunlit surface.*

81

3 Extend

LabBook   PG 88
Biome Business

DEMONSTRATION

Mock Permafrost Prepare for the demonstration by punching five holes in the bottom of two coffee cans and filling each can one-third full with potting soil. Slowly add water to one can until it begins to drain through the bottom; the soil should be moist but not saturated. Allow the excess water to drain, and pack the soil firmly. Place that can in a freezer for 6 to 8 hours. Bring the two cans to class, and have students gather at a sink. Hold the can with the unfrozen soil over the sink, and slowly pour a glass of water onto the soil. Repeat with the frozen can. Discuss with students why muddy or "marshy" areas form in the frozen soil and why the soil did not drain.

GROUP ACTIVITY

Have groups write a brochure for a summer camp in the biome of their choice. Suggest that they include information about the environment that will entice people to come and helpful tips about how to prepare for the area's climate.

WEIRD SCIENCE

Conical hills, called pingos, form in the arctic tundra when frozen ground water trapped under the permafrost is forced up due to pressure. As the frozen ground water rises, it pushes the ground over it upward. Pingos can be up to 46 m high and 400 m across.

MISCONCEPTION
///ALERT\\\

Students may think that the tundra is a relatively small and barren portion of the world. Actually, 1/10 of the Earth's land is tundra, and about 600 species of plants are native to the biome. Ninety-nine percent of those plants are perennials—the growing season is too short for annuals, which need time to produce flowers and seeds.

Quiz

1. What are the three major climate zones? (the tropical zone, the temperate zone, and the polar zone)

2. Can a climate zone contain more than one biome? (A climate zone may contain several different biomes.)

3. What is a microclimate? (a small region with unique climate characteristics)

ALTERNATIVE ASSESSMENT

Concept Mapping
Have students create a concept map that shows how each of the nine biomes is influenced by precipitation and temperature.

RESEARCH

Have students research the various microclimates that can be found on either side of Mount Shasta, in California. Have them find out which plants and animals are dominant at different elevations. They may want to present their findings in a color-coded diagram of the mountain with a key that explains the main characteristics of each microclimate.

PG 109

For the Birds

Reinforcement Worksheet
"A Tale of Three Climates"

Avg. Temperature Range:
−10°C–15°C (14°F–59°F)

Avg. Yearly Precipitation:
40–61 cm

Soil Characteristics: acidic soil

Vegetation: mosses, lichens, conifers

Animals: birds, rabbits, moose, elk, wolves, lynxes, and bears

Figure 25 *The taiga is the major source of wood for paper.*

Physics CONNECTION

Roof temperatures can get so hot that you can fry an egg on them! In a study of roofs on a sunny day when the air temperature was 13°C, scientists recorded roof temperatures ranging from 18°C to 61°C depending on color and material of the roof.

To find out more about microclimates, turn to page 109 of the LabBook.

Taiga (Northern Coniferous Forest)

Just south of the tundra lies the taiga biome. The taiga, as shown in **Figure 25**, has long, cold winters and short, warm summers. Like the tundra, the soil during the winter is frozen. The majority of the trees are evergreen needle-leaved trees called *conifers,* such as pine, spruce, and fir trees. The needles and bendable branches allow these trees to shed heavy snow before they can be damaged. Conifer needles contain acidic substances. When the needles die and fall to the soil, they make the soil acidic. Most plants cannot grow in acidic soil, and therefore the forest floor is bare except for some mosses and lichens.

Microclimates

You have learned the types of biomes that are found in each climate zone. But the climate and the biome of a particular place can also be influenced by local conditions. **Microclimates** are small regions with unique climatic characteristics. For example, elevation can affect an area's climate and therefore its biome. Tundra and taiga biomes exist in the Tropics on high mountains. How is this possible? Remember that as the elevation increases, the atmosphere loses its ability to absorb and hold thermal energy. This results in lower temperatures.

Cities are also microclimates. In a city, temperatures can be 1°C to 2°C warmer than the surrounding rural areas. This is because buildings and pavement made of dark materials absorb solar radiation instead of reflecting it. There is also less vegetation to take in the sun's rays. This absorption of the sun's rays by buildings and pavement heats the surrounding air and causes temperatures to rise.

SECTION REVIEW

1. Describe how tropical deserts and temperate deserts differ.

2. List and describe the three major climate zones.

3. **Inferring Conclusions** Rank each biome according to how suitable it would be for growing crops. Explain your reasoning.

▼ Answers to Section Review

1. Answers will vary. Sample answer: Tropical deserts are hot deserts, and temperate deserts are cold deserts. Winters in tropical deserts are usually mild, but temperate deserts often receive light snow during winter.

2. The three climate zones are the tropical zone, the temperate zone, and the polar zone. The tropical zone receives the most direct solar radiation; therefore the temperatures are generally hot. Temperatures in the temperate zone tend to be moderate. The temperate zone experiences seasonal variations, such as warm summers and cold winters. During winter, the temperatures in the polar zone stay below freezing. During the summer, temperatures remain cold but they can be above freezing.

3. Answers will vary.

Terms to Learn

ice age
global warming
greenhouse effect

What You'll Do

◆ Describe how the Earth's climate has changed over time.

◆ Summarize the different theories that attempt to explain why the Earth's climate has changed.

◆ Explain the greenhouse effect and its role in global warming.

Changes in Climate

As you know, the weather constantly changes—sometimes several times in one day. Saturday, your morning baseball game was canceled because of rain, but by that afternoon the sun was shining. Now think about the climate where you live. You probably haven't noticed a change in climate, because climates change slowly. What causes climates to change? Until recently, climatic changes were connected only to natural causes. However, studies indicate that human activities may have an influence on climatic change. In this section, you will learn how natural and human factors may influence climatic change.

Ice Ages

The geologic record indicates that the Earth's climate has been much colder than it is today. In fact, much of the Earth was covered by sheets of ice during certain periods. An **ice age** is a period during which ice collects in high latitudes and moves toward lower latitudes. Scientists have found evidence of many major ice ages throughout the Earth's geologic history. The most recent ice age began about 2 million years ago.

Figure 26 *During the last glacial period, which ended 10,000 years ago, the Great Lakes were covered by an enormous block of ice that was 1.5 km high.*

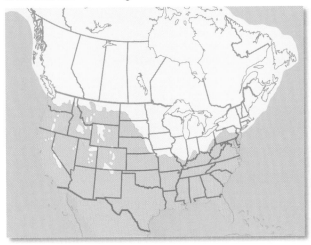

Glacial Periods During an ice age, there are periods of cold and periods of warmth. These periods are called glacial and interglacial periods. During *glacial periods*, the enormous sheets of ice advance, getting bigger and covering a larger area as shown in **Figure 26.** Because a large amount of ocean water is frozen during glacial periods, sea level drops.

83

Focus

Changes in Climate

This section describes significant changes in the Earth's climate. Students learn how different theories attempt to explain the cause of ice ages. Students also learn about the greenhouse effect and how the human production of greenhouse gases may contribute to global warming.

🔔 Bellringer

Have students imagine that the climate of the area where they live has changed so that it is now warmer than it used to be. Have students write down five different ways they think the area would be affected by warmer temperatures.

1 Motivate

DEMONSTRATION

The Greenhouse Effect Tell students that the glass windows in a greenhouse are similar to the Earth's atmosphere. The glass allows radiant energy to enter but prevents thermal energy from escaping. Have students place a thermometer in a plastic bag on a sunny windowsill. Place another thermometer next to the plastic bag. After 30 minutes, have a student read the two thermometers and compare the differences in temperature.
Sheltered English

 Directed Reading Worksheet Section 3

MISCONCEPTION ALERT

Students may be confused about the difference between an ice age and a glacial period. An ice age is the gradual cooling of the planet over thousands of years. During this time, glaciers repeatedly spread outward from the Earth's poles toward the equator. Ice ages are characterized by glacial periods (when glaciers spread) and interglacial periods (when glaciers retreat). Glacial periods can happen rather quickly—often in less than 30 years. Ice cores indicate that sudden glaciation periods could be caused by changes in major ocean currents or by volcanic eruptions.

MEETING INDIVIDUAL NEEDS

Advanced Learners Have students find out why scientists study the dust concentrations and gas composition of glacial ice in places such as Antarctica and Greenland. Have them explain why these and other data provide evidence about the last glacial period and other ice ages. Suggest that students share their findings with the class by giving an oral presentation.

In the early 1980s, astronomer Carl Sagan and four other scientists proposed a theory that the fallout from a nuclear war could result in disastrous climatic changes. The fallout, consisting of billions of tons of dust and ash ejected into the atmosphere, would act as a shield, blocking out so much of the sun's rays that it would cause periods of darkness and below-freezing temperatures possibly lasting a year or longer. This scenario is described as a "nuclear winter." Discuss with students the similarities between the nuclear-winter theory and the theory that suggests glaciation is caused by catastrophic events such as massive volcanic eruptions.

Teaching Transparency 179
"The Milankovitch Theory of the Causes of the Ice Ages"

Interglacial Periods   Warmer times that occur between glacial periods are called *interglacial periods*. During an interglacial period, the ice begins to melt and the sea level rises again. The last interglacial period began 10,000 years ago and is still occurring. Why do these periods occur? Will the Earth have another glacial period in the future? These questions have been debated by scientists for the past 200 years.

Motions of the Earth   There are many theories about the causes of ice ages. Each theory attempts to explain the gradual cooling that leads to the development of enormous ice sheets that periodically cover large areas of the Earth's surface. The *Milankovitch theory* explains why an ice age isn't just one long cold spell but instead alternates between cold and warm periods. Milutin Milankovitch, a Yugoslavian scientist, proposed that changes in the Earth's orbit and in the tilt of the Earth's axis cause ice ages, as illustrated in **Figure 27**.

Figure 27   *According to the Milankovitch theory, the amount of solar radiation the Earth receives varies due to three kinds of changes in the Earth's orbit.*

1 Over a period of 100,000 years, the Earth's orbit slowly changes from a more circular shape to a more elliptical shape. When the orbit is more elliptical, summers are hotter and winters are cooler. When the orbit is more circular, there is not as much seasonal change.

2 Over a period of 41,000 years, the tilt of the Earth's axis varies between 21.8° and 24.4°. When the tilt is at 24.4°, the poles receive more solar energy.

3 The Earth's axis traces a complete circle every 26,000 years. The circular motion of the Earth's axis determines the time of year that the Earth is closest to the sun.

✓ Self-Check

How do you think the Earth's elliptical orbit affects the amount of solar radiation that reaches the surface? *(See page 136 to check your answer.)*

Answer to Self-Check

The Earth's elliptical orbit causes seasonal differences. When the Earth's orbit is more elliptical, summers are hotter because the Earth is closer to the sun and receives more solar radiation. However, the Earth also moves farther from the sun in winter and receives less solar radiation. (Point out that the elliptical orbit of the Earth shown in **Figure 27** is exaggerated for effect. The change in orbits from circular to more elliptical is actually very slight.)

WEIRD SCIENCE

During the last glacial period, animals that live in the northern part of North America, such as the Arctic fox, wolf, grizzly bear, and caribou, were living in Oklahoma, Missouri, and Texas!

Volcanic Eruptions There are many natural factors that can affect global climate. Catastrophic events, such as volcanic eruptions, can influence climate. Volcanic eruptions send large amounts of dust, ash, and smoke into the atmosphere. Once in the atmosphere, the dust, smoke, and ash particles act as a shield, blocking out so much of the sun's rays that the Earth cools. **Figure 28** shows how dust particles from a volcanic eruption block the sun.

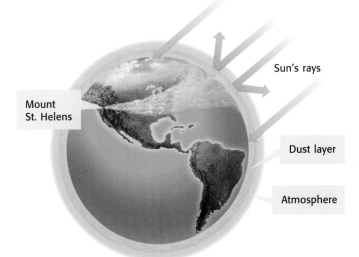

Sun's rays

Mount St. Helens

Dust layer

Atmosphere

Figure 28 *Volcanic eruptions, such as the one that occurred at Mount St. Helens, shown above, produce dust that reflects sunlight, as shown at left.*

Plate Tectonics The Earth's climate is further influenced by plate tectonics and continental drift. One theory proposes that ice ages occur when the continents are positioned closer to the polar regions. For example, approximately 250 million years ago, all the continents were connected near the South Pole in one giant landmass called Pangaea, as shown in **Figure 29.** During this time, ice covered a large area of the Earth's surface. As Pangaea broke apart, the continents moved toward the equator, and the ice age ended. During the last ice age, many large landmasses were positioned in the polar zones. Antarctica, northern North America, Europe, and Asia all were covered with large sheets of ice.

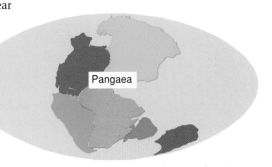

Pangaea

Figure 29 *Much of Pangaea—the part that is now Africa, South America, India, Antarctica, Australia, and Saudi Arabia—was covered by continental ice sheets.*

85

REAL-WORLD CONNECTION

Tell students that washing clothes in cold water instead of hot water can reduce the amount of carbon dioxide released into the atmosphere. This is because fossil fuels are used to heat the water. For example, a household that uses cold water to do two loads of laundry a week releases about 225 kg *less* carbon dioxide into the atmosphere each year. Have students calculate what the annual reduction in released carbon dioxide would be if the family of every student in the class used cold water for laundry.

CROSS-DISCIPLINARY FOCUS

History Archaeological evidence suggests that North America was originally settled by people who walked here from Asia between 15,000 and 25,000 years ago. At that time, sea levels were as much as 150 m lower, exposing the land between Alaska and Siberia. The sea level was lower because the ice sheets that spread across the Northern Hemisphere contained vast amounts of water. Have students write a short report about other human migrations thought to be caused by climate change.

Multicultural CONNECTION

It is difficult for scientists to predict climate changes because weather data has been accurately recorded for less than 200 years. However, the meteorological records of the Chinese date back to 1216 B.C. While these records do not indicate temperature, they do record rainfall, sleet, snow, humidity, and wind direction. There are also comments on unusually warm or cool temperatures. Have interested students research the reasons for collecting the data and find out how it is used today.

3 Extend

CONNECT TO LIFE SCIENCE

There are many consequences of global warming. One possibility is the spread of tropical diseases, such as malaria and dengue fever. Both of these diseases are carried by a specific species of mosquito. These mosquitoes have a minimum temperature at which they can survive and breed. Ask students to create a temperature map for the world and identify the areas that are most likely to have a problem with these diseases.

LabBook **PG 108**
Global Impact

 Critical Thinking Worksheet
"Cyberspace Heats Up"

Science Skills Worksheet
"Understanding Bias"

÷ 5 ÷ Ω ≤ ∞ +Ω √ 9 ∞ ≤ Σ 2
+ ÷

MATH BREAK

The Ride to School

Find out how much carbon dioxide is released into the atmosphere each month from the car or bus that transports you to school.

1. Figure out the distance from your home to school.
2. From this figure, calculate how many kilometers you travel to and from school, in a car or bus, per month.
3. Divide this number by 20. This represents approximately how many gallons of gas are used during your trips to school.
4. If burning 1 gal of gasoline produces 9 kg of carbon dioxide, how much carbon dioxide is released?

 internet connect

SCI LINKS
NSTA

TOPIC: Changes in Climate
GO TO: www.scilinks.org
*sci*LINKS NUMBER: HSTE415

TOPIC: Modeling Earth's Climate
GO TO: www.scilinks.org
*sci*LINKS NUMBER: HSTE420

Global Warming

Is the Earth really experiencing global warming? **Global warming** is a rise in average global temperatures that can result from an increase in the greenhouse effect. To understand how global warming works, you must first learn about the greenhouse effect.

Greenhouse Effect The **greenhouse effect** is the Earth's natural heating process, in which gases in the atmosphere trap thermal energy. The Earth's atmosphere performs the same function as the glass windows in a car. Think about the car illustrated in **Figure 30**. It's a hot summer day, and you are about to get inside the car. You immediately notice that it feels hotter inside the car than outside. Then you sit down and—ouch!—you burn yourself on the seat.

Figure 30 *Sunlight streams into the car through the clear glass windows. The seats absorb the radiant energy and change it into thermal energy. The energy is then trapped in the car.*

Window to the World Greenhouse gases allow sunlight to pass through the atmosphere. It is absorbed by the Earth's surface and reradiated as thermal energy. Many scientists hypothesize that the rise in global temperatures is due to an increase of carbon dioxide, a greenhouse gas, as a result of human activity. Most evidence indicates that the increase in carbon dioxide is caused by the burning of fossil fuels that releases carbon dioxide into the atmosphere.

SCIENTISTS AT ODDS

Svante Arrhenius was the first scientist to propose that the global climate was changing as a result of human activities. He put forth his theory in 1905. Other scientists ridiculed his ideas. It was not until much later that scientists took a serious look at the effects of carbon dioxide on the global climate.

Another factor that may add to global warming is deforestation. *Deforestation* is the process of clearing forests, as shown in **Figure 31**. In many countries around the world, forests are being burned to clear land for agriculture. All types of burning release carbon dioxide into the atmosphere, thereby increasing the greenhouse effect. Plants use carbon dioxide to make food. As plants are removed from the Earth, the carbon dioxide that would have been used by the plants builds up in the atmosphere.

Figure 31 *Clearing land by burning leads to increased levels of carbon dioxide in the atmosphere.*

Consequences of Global Warming Many scientists think that if the average global temperature continues to rise, some regions of the world might experience flooding. Warmer temperatures could cause the icecaps to melt, raising the sea level and flooding low-lying areas, such as the coasts.

Areas that receive little rainfall, such as deserts, might receive even less due to increased evaporation. Scientists predict that the Midwest, an agricultural area, could experience warmer, drier conditions. A change in climate such as this could harm crops. But farther north, such as in Canada, weather conditions for farming would improve.

Reducing Pollution

A city just received a warning from the Environmental Protection Agency for exceeding the automobile fuel emissions standards. If you were the city manager, what suggestions would you make to reduce the amount of automobile emissions?

SECTION REVIEW

1. How has the Earth's climate changed over time? What might have caused these changes?

2. Explain how the greenhouse effect warms the Earth.

3. What are two ways that humans contribute to the increase in carbon dioxide levels in the atmosphere?

4. **Analyzing Relationships** How will the warming of the Earth affect agriculture in different parts of the world?

internet**connect**

SC*i*LINKS

NSTA

TOPIC: Changes in Climate
GO TO: www.scilinks.org
*sci*LINKS **NUMBER:** HSTE415

87

▼ *Answers to Section Review*

1. The Earth's climate has been both warmer and colder than it is today. Climate change may result from variations in the Earth's orbit; catastrophic events, such as volcanic eruptions; and the movement of the continents by plate tectonics and continental drift.

2. Greenhouse gases allow sunlight to pass through the atmosphere, where it is absorbed by the Earth's surface and reradiated as thermal energy. The greenhouse gases absorb the thermal energy as it moves out of the atmosphere.

3. burning fossil fuels and deforestation

4. Answers will vary. The warming of the Earth would change the climate. Areas that were suitable for farming might become warmer and drier.

Biome Business
Teacher's Notes

Time Required

One 45-minute class period

Lab Ratings

EASY ⟶ HARD

TEACHER PREP ⚗
STUDENT SET-UP ⚗
CONCEPT LEVEL ⚗⚗⚗
CLEAN UP ⚗

MATERIALS

Student groups will need a general map to identify their biome location.

Preparation Notes

Remind students not to use seasonal terms such as spring and fall because some of the biomes in the Southern Hemisphere may experience seasons opposite those of the Northern Hemisphere.

 Datasheets for LabBook

David Sparks
Redwater Jr. High
Redwater, Texas

Biome Business

You have just been hired as an assistant to a world-famous botanist. You have been provided with climatographs for three biomes. A *climatograph* is a graph that shows the temperature and precipitation patterns for an area for a year.

You can use the information provided in the graphs to determine the type of climate in each biome. You also have a general map of the biomes, but nothing is labeled. Using this information, you must figure out what the environment will be like in each biome.

In this activity, you will use climatographs and maps to determine where you will be traveling. You can find the exact locations by tracing the general maps and matching them to the map at the bottom of the page.

Procedure TRY at HOME

1. Look at each climatograph. The shaded areas show the average precipitation for the biome. The red line shows the average temperature.

2. Use the climatographs to determine the climate patterns for each biome. Compare the maps with the biome map on page 74 to find the exact location of each region.

Analysis

3. Describe the precipitation patterns of each biome by answering the following questions:
   a. When does it rain the most in this biome?
   b. Do you think the biome is relatively dry, or do you think it rains a lot?

4. Describe the temperature patterns of each biome by answering the following questions:
   a. What are the warmest months of the year?
   b. Does the biome seem to have temperature cycles, like seasons, or is the temperature almost always the same?
   c. Do you think the biome is warm or cool? Explain your answer.

5. Name each biome.

6. Where is each biome located?

88

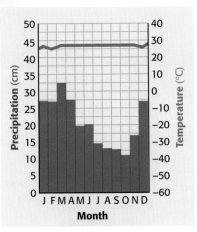

Biome A

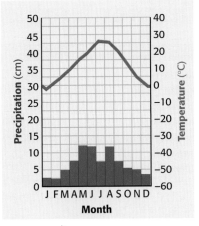

Biome C

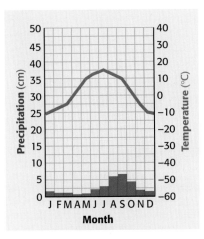

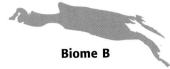

Biome B

Going Further

In a cardboard box no bigger than a shoe box, build a model of one of the biomes that you investigated. Include things to represent the biome, such as the plants and animals that inhabit the area. Use magazines, photographs, colored pencils, plastic figurines, clay, or whatever you like. Be creative!

Answers

3. a. In Biome A, the rain falls throughout the year but is heaviest in March.
   In Biome B, the rain is heaviest in September.
   In Biome C, the rain is heaviest in May, June, and August.

   b. Biome A is very rainy and wet. Biomes B and C are relatively dry, but some months are rainier than others.

4. a. Biome A has a relatively constant temperature throughout the year. Biomes B and C experience their warmest months from June to August.

   b. Biome A has a constant temperature throughout the year. Biomes B and C experience seasonal cycles.

   c. Biome A is warm and the temperature is high year-round. Biome B has a cooler climate and the climatograph shows cooler temperatures year-round. Biome C has a moderate climate in the early and late months of the year, but the temperature is quite hot in the middle months of the year.

5. Biome A = a tropical rain forest
   Biome B = a taiga
   Biome C = a temperate grassland

6. Biome A = the west coast of Africa near the equator
   Biome B = northern Asia
   Biome C = midwestern United States

Chapter Highlights

Chapter Highlights

SECTION 1

weather the condition of the atmosphere at a particular time and place

climate the average weather conditions in an area over a long period of time

latitude the distance north or south from the equator; measured in degrees

prevailing winds winds that blow mainly from one direction

elevation the height of an object above sea level; the height of surface landforms above sea level

surface currents a streamlike movement of water that occurs at or near the surface of the ocean

SECTION 1

Vocabulary

- **weather** *(p. 68)*
- **climate** *(p. 68)*
- **latitude** *(p. 69)*
- **prevailing winds** *(p. 71)*
- **elevation** *(p. 72)*
- **surface currents** *(p. 73)*

Section Notes

- Weather is the condition of the atmosphere at a particular time and place. Climate is the average weather conditions in a certain area over a long period of time.

- Climate is determined by temperature and precipitation.

- Climate is controlled by factors such as latitude, elevation, wind patterns, local geography, and ocean surface currents.

- The amount of solar energy an area receives is determined by the area's latitude.

- The seasons are a result of the tilt of the Earth's axis and its path around the sun.

- The amount of moisture carried by prevailing winds affects the amount of precipitation that falls.

- As elevation increases, temperature decreases.

- Mountains affect the distribution of precipitation. The dry side of the mountain is called the rain shadow.

- As ocean surface currents move across the Earth, they redistribute warm and cool water. The temperature of the surface water affects the air temperature.

☑ Skills Check

Visual Understanding

THE SEASONS Seasons are determined by latitude. The diagram on page 70 shows how the tilt of the Earth affects how much solar energy an area receives as the Earth moves around the sun.

THE RAIN SHADOW The illustration on page 72 shows how the climates on two sides of a mountain can be very different. A mountain can affect the climate of areas nearby by influencing the amount of precipitation these areas receive.

LAND BIOMES OF THE EARTH Look back at Figure 10 on page 74 to review the distribution of the Earth's Land Biomes.

Lab and Activity Highlights

Biome Business `PG 88`

Global Impact `PG 108`

For the Birds `PG 109`

Datasheets for LabBook
(blackline masters for these labs)

SECTION 2

Vocabulary

biome *(p. 74)*

tropical zone *(p. 75)*

temperate zone *(p. 78)*

deciduous *(p. 78)*

evergreens *(p. 78)*

polar zone *(p. 80)*

microclimate *(p. 82)*

Section Notes

- The Earth is divided into three climate zones according to latitude—the tropical zone, the temperate zone, and the polar zone.

- The tropical zone is the zone around the equator. The tropical rain forest, tropical desert, and tropical savanna are in this zone.

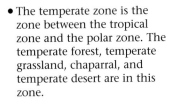

- The temperate zone is the zone between the tropical zone and the polar zone. The temperate forest, temperate grassland, chaparral, and temperate desert are in this zone.

- The polar zones are the northernmost and southernmost zones. The taiga and tundra are in this zone.

Labs

For the Birds *(p. 109)*

SECTION 3

Vocabulary

ice age *(p. 83)*

global warming *(p. 86)*

greenhouse effect *(p. 86)*

Section Notes

- Explanations for the occurrence of ice ages include changes in the Earth's orbit, volcanic eruptions, and plate tectonics and continental drift.

- Some scientists believe that global warming is occurring as a result of an increase in carbon dioxide from human activity.

- If global warming continues, it could drastically change climates, causing either floods or drought.

Labs

Global Impact *(p. 108)*

VOCABULARY DEFINITIONS, continued

SECTION 2

biome a large region characterized by a specific type of climate and the plants and animals that live there

tropical zone the warm zone located around the equator

temperate zone the climate zone between the Tropics and the polar zone

deciduous describes trees that lose their leaves when the weather becomes cold

evergreens trees that keep their leaves year-round

polar zone the northernmost and southernmost climate zones

microclimate a small region with unique climatic characteristics

SECTION 3

ice age a period during which ice collects in high latitudes and moves toward lower latitudes

global warming a rise in average global temperatures

greenhouse effect the natural heating process of a planet, such as the Earth, by which gases in the atmosphere trap thermal energy

Vocabulary Review Worksheet

Blackline masters of these Chapter Highlights can be found in the **Study Guide.**

---

 internetconnect

go hrw .com

GO TO: go.hrw.com

Visit the **HRW** Web site for a variety of learning tools related to this chapter. Just type in the keyword:

KEYWORD: HSTCLM

SCiLINKS

N S T A

GO TO: www.scilinks.org

Visit the **National Science Teachers Association** on-line Web site for Internet resources related to this chapter. Just type in the *sci*LINKS number for more information about the topic:

TOPIC: What Is Climate?	*sci*LINKS NUMBER: HSTE405
TOPIC: Climates of the World	*sci*LINKS NUMBER: HSTE410
TOPIC: Changes in Climate	*sci*LINKS NUMBER: HSTE415
TOPIC: Modeling Earth's Climate	*sci*LINKS NUMBER: HSTE420

91

Lab and Activity Highlights

LabBank

 Whiz-Bang Demonstrations,
How Humid Is It?

 Calculator-Based Labs,
- What Causes the Seasons?
- Heating of Land and Water

Long-Term Projects & Research Ideas,
Sun-Starved in Fairbanks

Chapter Review
Answers

USING VOCABULARY

1. Climate
2. Latitude
3. tropical
4. deciduous
5. tundra
6. global warming

UNDERSTANDING CONCEPTS

Multiple Choice

7. b
8. b
9. a
10. b
11. c
12. a
13. d
14. b

Short Answer

15. Higher latitudes receive less solar radiation because the sun's rays strike the Earth's surface at a less direct angle. This spreads the same amount of solar energy over a larger area, resulting in lower temperatures.

16. The amount of precipitation an area receives can depend on whether the region's prevailing winds form from a warm air mass or from a cold air mass. If the winds form from a warm air mass, they will probably carry moisture. If the winds form from a cold air mass, they will probably be dry. Precipitation is more likely to occur when the prevailing winds are warm and moist.

17. Answers will vary. Sample answer: Tundra and taiga biomes can be found on tropical mountains. This is because air at higher elevations retains less thermal energy than air at lower elevations.

18. Plants have adapted by developing fleshy leaves to store water and a waxy coating to prevent water loss. Animals are more active at night, when temperatures are cooler, and they burrow during the day.

Chapter Review

USING VOCABULARY

To complete the following sentences, choose the correct term from each pair of terms listed below.

1. __?__ is the condition of the atmosphere in a certain area over a long period of time. *(Weather or Climate)*

2. __?__ is the distance north and south from the equator measured in degrees. *(Longitude or Latitude)*

3. Savannas are grasslands located in the __?__ zone between 23.5° north latitude and 23.5° south latitude. *(temperate or tropical)*

4. Trees that lose their leaves are found in a(n) __?__ forest. *(deciduous or evergreen)*

5. Frozen land in the polar zone is most often found in a __?__. *(taiga or tundra)*

6. A rise in global temperatures due to an increase in carbon dioxide is called __?__. *(global warming or the greenhouse effect)*

UNDERSTANDING CONCEPTS

Multiple Choice

7. The tilt of Earth as it orbits the sun causes
   a. global warming.
   b. different seasons.
   c. a rain shadow.
   d. the greenhouse effect.

8. What factor affects the prevailing winds as they blow across a continent, producing different climates?
   a. latitude
   b. mountains
   c. forests
   d. glaciers

9. What factor determines the amount of solar energy an area receives?
   a. latitude
   b. wind patterns
   c. mountains
   d. ocean currents

10. What climate zone has the coldest average temperature?
    a. tropical
    b. polar
    c. temperate
    d. tundra

11. What biome is not located in the tropical zone?
    a. rain forest
    b. savanna
    c. chaparral
    d. desert

12. What biome contains the greatest number of plant and animal species?
    a. rain forest
    b. temperate forest
    c. grassland
    d. tundra

13. Which of the following is not a theory for the cause of ice ages?
    a. the Milankovitch theory
    b. volcanic eruptions
    c. plate tectonics
    d. the greenhouse effect

14. Which of the following is thought to contribute to global warming?
    a. wind patterns
    b. deforestation
    c. ocean surface currents
    d. microclimates

19. Both the tundra and desert biomes receive very little precipitation.

Concept Mapping

20. An answer to this exercise can be found at the front of this book.

Concept Mapping
Transparency 17

Short Answer

15. Why do higher latitudes receive less solar radiation than lower latitudes?

16. How does wind influence precipitation patterns?

17. Give an example of a microclimate. What causes the unique temperature and precipitation characteristics of this area?

18. How have desert plants and animals adapted to this biome?

19. How are tundra and deserts similar?

Concept Mapping

20. Use the following terms to create a concept map: climate, global warming, deforestation, greenhouse effect, flooding.

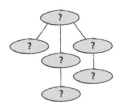

CRITICAL THINKING AND PROBLEM SOLVING

Write one or two sentences to answer the following questions:

21. Explain how ocean surface currents are responsible for milder climates.

22. In your own words, explain how a change in the Earth's orbit can affect the Earth's climates as proposed by Milutin Milankovitch.

23. Explain why the climate differs drastically on each side of the Rocky Mountains.

24. What are some steps you and your family can take to reduce the amount of carbon dioxide that is released into the atmosphere?

MATH IN SCIENCE

25. If the air temperature near the shore of a lake measures 24°C, and if the temperature increases by 0.05°C every 10 m traveled away from the lake, what would the air temperature be 1 km from the lake?

INTERPRETING GRAPHICS

The following illustration shows the Earth's orbit around the sun.

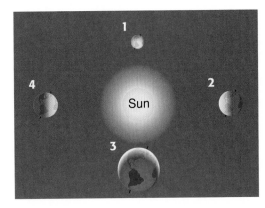

26. At what position, **1**, **2**, **3**, or **4**, is it spring in the Southern Hemisphere?

27. At what position does the South Pole receive almost 24 hours of daylight?

28. Explain what is happening in each climate zone in both the Northern Hemisphere and Southern Hemisphere at position **4**.

Reading Check-up

Take a minute to review your answers to the Pre-Reading Questions found at the bottom of page 66. Have your answers changed? If necessary, revise your answers based on what you have learned since you began this chapter.

93

CRITICAL THINKING AND PROBLEM SOLVING

21. Warmer surface currents heat the surrounding air, and colder surface currents cool the surrounding air. A warm surface current might bring warmer temperatures to a cold area. A cold surface current might cool an area that is generally hot.

22. Answers will vary. Sample answer: A change in the Earth's orbit can affect the Earth's climates by limiting or increasing the amount of solar radiation the Earth receives.

23. The climate differs on each side of the Rocky Mountains because the mountains affect the distribution of precipitation. The windward side receives more precipitation because as the warm air is forced to rise, it releases precipitation. As the dry air crosses the mountain, it sinks, warming and absorbing the moisture.

24. Answers will vary.

MATH IN SCIENCE

25. 1 km = 1,000 m
1,000 m ÷ 10 m = 100
100 × .05°C = 5°C
24°C + 5°C = 29°C

INTERPRETING GRAPHICS

26. 3

27. 2

28. In the tropical zone temperatures are warm. The temperate zone in the Northern Hemisphere is experiencing summer. The temperate zone in the Southern Hemisphere is experiencing winter. Deciduous trees have shed their leaves. The polar zone in the Northern Hemisphere is experiencing almost 24 hours of daylight. Temperatures are cool, and the top meter of soil is thawing. The polar zone in the Southern Hemisphere is experiencing almost 24 hours of night. Temperatures are extremely cold, and the soil is frozen.

Blackline masters of this Chapter Review can be found in the **Study Guide.**

ACROSS THE SCIENCES

Background

Another pattern in the global weather system is called the *southern oscillation*. In 1924, the British mathematician Gilbert Walker described a see-saw pattern in tropical Pacific air pressure. He found that when air pressure is low around Australia, it is high to the east, in Tahiti. Conversely, when air pressure is high in Australia, it is low in Tahiti. This see-saw effect is called the southern oscillation.

By the 1970s, scientists realized that El Niño and the southern oscillation are part of a huge oceanic-atmospheric system that affects the weather in many parts of the world.

Blame "The Child"

El Niño, which is Spanish for "the child," is the name of a weather event that occurs in the Pacific Ocean. Every 2 to 12 years, the interaction between the ocean surface and atmospheric winds creates El Niño. This event influences weather patterns in many regions of the world.

Difficult Breathing

For Indonesia and Malaysia, El Niño meant droughts and forest fires in 1998. Thousands of people in these countries suffered from respiratory ailments from breathing the smoke caused by these fires. Heavy rains in San Francisco created extremely high mold-spore counts. These spores cause problems for people with allergies. The spore count in February in San Francisco is usually between 0 and 100. In 1998, the count was often higher than 8,000!

Rodent Invasion

In areas where El Niño creates heavy rains, the result is lush vegetation. This lush vegetation provides even more food and shelter for rodents. As the rodent population increases, so does the threat of the diseases they spread. In states like Arizona, Colorado, and New Mexico, this means there is a greater chance among humans of contracting hantaviral pulmonary syndrome (HPS).

HPS is carried by deer mice and remains in their urine and feces. People are infected when they inhale dust contaminated with mouse feces or urine. Once infected, a person experiences flulike symptoms that can sometimes lead to fatal kidney or lung disease.

More Rodents and Insects

Heavy rains near Los Angeles might encourage a rodent-population explosion in the mountains east of the city. If so, there could be an increase in the number of rodents infected with bubonic plague. More infected rodents means more infected fleas, which carry bubonic plague to humans.

Ticks and mosquitoes could also increase in number. These insects can spread disease too. For example, ticks can carry Lyme disease, ehrlichiosis, babesiosis, and Rocky Mountain spotted fever. Mosquitoes can spread malaria, dengue fever, encephalitis, and Rift Valley fever.

◀ *If this flea carries bubonic plague bacteria, just one bite can infect a person.*

What About Camping?

Because all of these diseases can be fatal to humans, people must take precautions. Camping in the great outdoors increases the risk of infection. Campers should steer clear of rodents and their burrows. Don't forget to dust family pets with flea powder, and don't let them roam free. Try to remember that an ounce of prevention is worth a pound of cure.

Find Out More

▶ How do you think El Niño affects the fish and mammals that live in the ocean? Write your answer in your ScienceLog, and then do some research to see if you are correct.

94

Answer to Find Out More

Answers will vary. Due to a shortage of phytoplankton, El Niño forces fish to find food in other areas. For example, mako sharks, which normally live in warm tropical waters, were found in the chilly waters of Monterey Bay, California. Birds such as pelicans might end up in areas that are far from their normal habitat. This was the case in Arica, Chile, in August 1997, when the number of local pelicans grew from 200 to 4,000 in just a few weeks. The population surge was believed to be caused by El Niño. If food is not available, El Niño can cause starvation among some species. The Galápagos penguin population, for example, decreased by 50 percent due to weather caused by El Niño and La Niña.

Science, Technology, and Society

Some Say Fire, Some Say Ice . . .

The Earth's climate has undergone many drastic changes. For example, 6,000 years ago in the part of North Africa that is now a desert, hippos, crocodiles, and early Stone Age people shared shallow lakes that covered the area. Grasslands stretched as far as the eye could see.

Scientists have known for many years that Earth's climate has changed. What they didn't know was why. Using supercomputers and complex computer programs, scientists may now be able to explain why North Africa's lakes and grasslands became a desert. And that information may be useful for predicting future heat waves and ice ages.

Climate Models

Scientists who study Earth's atmosphere have developed climate models to try to imitate Earth's climate. A climate model is like a very complicated recipe with thousands of ingredients. These models do not make exact predictions about future climates, but they do estimate what might happen.

What ingredients are included in a climate model? One important ingredient is the level of greenhouse gases (especially carbon dioxide) in the atmosphere. Land and ocean water temperatures from around the globe are other ingredients. So is information about clouds, cloud cover, snow, and ice cover. And in more recent models, scientists have included information about ocean currents.

A Challenge to Scientists

Earth's atmosphere-ocean climate system is extremely complex. One challenge for scientists is to understand all the system's parts. Another is to understand how those parts work together. But understanding Earth's climate system is critical. An accurate climate model should help scientists predict heat waves, floods, and droughts.

Even the best available climate models must be improved. The more information scientists can include in a climate model, the more accurate the results. Today data are available from more locations, and scientists need more-powerful computers to process all the data.

As more-powerful computers are developed to handle all the data in a climate model, scientists' understanding of Earth's climate changes will improve. This knowledge should help scientists better predict the impact human activities have on global climate. And these models could help scientists prevent some of the worst effects of climate change, such as global warming or another ice age.

▲ *This meteorologist is using a high-powered supercomputer to do climate modeling.*

A Challenge for You

▶ Earth's oceans are a major part of the climate model. Find out some of the ways oceans affect climate. Do you think human activities are changing the oceans?

95

SCIENCE, TECHNOLOGY, AND SOCIETY

Some Say Fire, Some Say Ice . . .

Teaching Strategy

There are many sites on the Internet that contain information about global and regional climate models. A search using the keywords "climate models" should produce plenty of information for you and your students.

Answers to A Challenge for You

Student answers will vary. Sample answer: The circulation of warm and cold surface currents affects climates. Students may mention El Niño and La Niña, or the absorption of carbon dioxide by ocean phytoplankton. Students may also note that the ocean provides enormous amounts of water for the water cycle.

SAFETY FIRST!

Exploring, inventing, and investigating are essential to the study of science. However, these activities can also be dangerous. To make sure that your experiments and explorations are safe, you must be aware of a variety of safety guidelines.

You have probably heard of the saying, "It is better to be safe than sorry." This is particularly true in a science classroom where experiments and explorations are being performed. Being uninformed and careless can result in serious injuries. Don't take chances with your own safety or with anyone else's.

Following are important guidelines for staying safe in the science classroom. Your teacher may also have safety guidelines and tips that are specific to your classroom and laboratory. Take the time to be safe.

Safety Rules!

Start Out Right

Always get your teacher's permission before attempting any laboratory exploration. Read the procedures carefully, and pay particular attention to safety information and caution statements. If you are unsure about what a safety symbol means, look it up or ask your teacher. You cannot be too careful when it comes to safety. If an accident does occur, inform your teacher immediately, regardless of how minor you think the accident is.

Safety Symbols

All of the experiments and investigations in this book and their related worksheets include important safety symbols to alert you to particular safety concerns. Become familiar with these symbols so that when you see them, you will know what they mean and what to do. It is important that you read this entire safety section to learn about specific dangers in the laboratory.

If you are instructed to note the odor of a substance, wave the fumes toward your nose with your hand. Never put your nose close to the source.

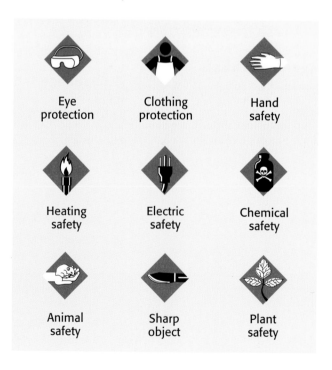

Eye protection

Clothing protection

Hand safety

Heating safety

Electric safety

Chemical safety

Animal safety

Sharp object

Plant safety

Eye Safety

Wear safety goggles when working around chemicals, acids, bases, or any type of flame or heating device. Wear safety goggles any time there is even the slightest chance that harm could come to your eyes. If any substance gets into your eyes, notify your teacher immediately, and flush your eyes with running water for at least 15 minutes. Treat any unknown chemical as if it were a dangerous chemical. Never look directly into the sun. Doing so could cause permanent blindness.

Avoid wearing contact lenses in a laboratory situation. Even if you are wearing safety goggles, chemicals can get between the contact lenses and your eyes. If your doctor requires that you wear contact lenses instead of glasses, wear eye-cup safety goggles in the lab.

Safety Equipment

Know the locations of the nearest fire alarms and any other safety equipment, such as fire blankets and eyewash fountains, as identified by your teacher, and know the procedures for using them.

Be extra careful when using any glassware. When adding a heavy object to a graduated cylinder, tilt the cylinder so the object slides slowly to the bottom.

Neatness

Keep your work area free of all unnecessary books and papers. Tie back long hair, and secure loose sleeves or other loose articles of clothing, such as ties and bows. Remove dangling jewelry. Don't wear open-toed shoes or sandals in the laboratory. Never eat, drink, or apply cosmetics in a laboratory setting. Food, drink, and cosmetics can easily become contaminated with dangerous materials.

Certain hair products (such as aerosol hair spray) are flammable and should not be worn while working near an open flame. Avoid wearing hair spray or hair gel on lab days.

Sharp/Pointed Objects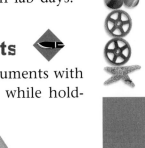

Use knives and other sharp instruments with extreme care. Never cut objects while holding them in your hands. Place objects on a suitable work surface for cutting.

Heat

Wear safety goggles when using a heating device or a flame. Whenever possible, use an electric hot plate as a heat source instead of an open flame. When heating materials in a test tube, always angle the test tube away from yourself and others. In order to avoid burns, wear heat-resistant gloves whenever instructed to do so.

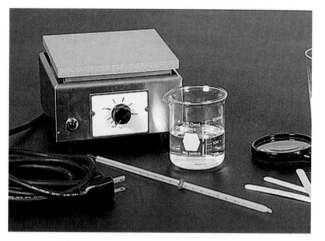

Electricity

Be careful with electrical cords. When using a microscope with a lamp, do not place the cord where it could trip someone. Do not let cords hang over a table edge in a way that could cause equipment to fall if the cord is accidentally pulled. Do not use equipment with damaged cords. Be sure your hands are dry and that the electrical equipment is in the "off" position before plugging it in. Turn off and unplug electrical equipment when you are finished.

Chemicals

Wear safety goggles when handling any potentially dangerous chemicals, acids, or bases. If a chemical is unknown, handle it as you would a dangerous chemical. Wear an apron and safety gloves when working with acids or bases or whenever you are told to do so. If a spill gets on your skin or clothing, rinse it off immediately with water for at least 5 minutes while calling to your teacher.

Never mix chemicals unless your teacher tells you to do so. Never taste, touch, or smell chemicals unless you are specifically directed to do so. Before working with a flammable liquid or gas, check for the presence of any source of flame, spark, or heat.

Animal Safety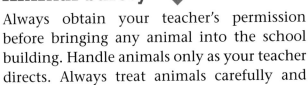

Always obtain your teacher's permission before bringing any animal into the school building. Handle animals only as your teacher directs. Always treat animals carefully and with respect. Wash your hands thoroughly after handling any animal.

Plant Safety

Do not eat any part of a plant or plant seed used in the laboratory. Wash hands thoroughly after handling any part of a plant. When in nature, do not pick any wild plants unless your teacher instructs you to do so.

Glassware

Examine all glassware before use. Be sure that glassware is clean and free of chips and cracks. Report damaged glassware to your teacher. Glass containers used for heating should be made of heat-resistant glass.

Boiling Over!
Teacher's Notes

Time Required

One 45-minute class period

Lab Ratings

EASY ——————————→ HARD

   TEACHER PREP 🝙

   STUDENT SET-UP 🝙🝙🝙

   CONCEPT LEVEL 🝙🝙

   CLEAN UP 🝙🝙

MATERIALS

The materials listed on the student page are enough for a group of 3–4 students.

Safety Caution

Remind students to review all safety cautions and icons before beginning this lab activity.

Preparation Notes

Begin the activity by leading a discussion of how thermometers work. Have students observe a regular thermometer. Ask students what parts are involved to make a thermometer work. (a receptacle containing the fluid referred to as the bulb, a tube, and air in the tube)

 Datasheets for LabBook

Boiling Over!

Safety Industries, Inc., would like to offer the public safer alternatives to the mercury thermometer. Many communities have complained that the glass thermometers are easy to break, and people are concerned about mercury poisoning. As a result, we would like your team of inventors to come up with a workable prototype that uses water instead of mercury. Safety Industries would like to offer a contract to the team that comes up with the best substitute for a mercury thermometer. In this activity, you will design and test your own water thermometer. Good luck!

Ask a Question

1. What conditions cause the liquid to rise in a thermometer? How can I use this information to build a thermometer?

Form a Hypothesis

2. Brainstorm with a classmate to design a thermometer that requires only water. Sketch your design in your ScienceLog. Write a one-sentence hypothesis that describes how your thermometer will work.

Test the Hypothesis

3. Follow your design to build a thermometer using only materials from the materials list. Like a mercury thermometer, your thermometer will need a bulb and a tube. However, the liquid in your thermometer will be water.

4. To test your design, place the aluminum pie pan on a hot plate. Carefully pour water into the pan until it is halfway full. Allow the water to heat.

5. Put on your gloves, and carefully place the "bulb" of your thermometer in the hot water. Observe the water level in the tube. Does it rise?

6. If the water level does not rise, adjust your design as necessary, and repeat steps 3–5. When the water level does rise, sketch your final design in your ScienceLog.

7. After you finalize your design, you must calibrate your thermometer with a laboratory thermometer by taping an index card to the thermometer tube so that the entire part of the tube protruding from the "bulb" of the thermometer touches the card.

Materials

- heat-resistant gloves
- aluminum pie pan
- hot plate
- water
- assorted containers, such as plastic bottles, soda cans, film canisters, medicine bottles, test tubes, balloons, and yogurt containers with lids
- assorted tubes, such as clear inflexible plastic straws or 5 mm diameter plastic tubing, 30 cm long
- modeling clay
- food coloring
- pitcher
- transparent tape
- index card
- Celsius thermometer
- a paper cone-shaped filter or funnel
- 2 large plastic-foam cups
- ice cubes
- metric ruler

Daniel Bugenhagen
Yutan Jr.–Sr. High
Yutan, Nebraska

8. Place the cone-shaped filter or funnel into the plastic-foam cup. Carefully pour hot water from the hot plate into the filter or funnel. Be sure that no water splashes or spills.

9. Place your own thermometer and a laboratory thermometer in the hot water. Mark the water level on the index card as it rises. Observe and record the temperature on the laboratory thermometer, and write this value on the card beside the mark.

10. Repeat steps 8–9 with warm water from the faucet.

11. Repeat steps 8–9 with ice water.

12. Divide the markings on the index card into equally sized increments, and write the corresponding temperatures on the index card.

Analyze the Results

13. How effective is your thermometer at measuring temperature?

14. Compare your thermometer design with other students' designs. How would you modify your design to make your thermometer measure temperature even better?

Draw Conclusions

15. Take a class vote to see which design should be chosen for a contract with Safety Industries. Why was this thermometer chosen? How did it differ from other designs in the class?

General Design

A water thermometer has a receptacle containing water and air with a tube protruding from the receptacle. A trick to getting the water thermometer to work well is to allow a lot of air in the "bulb" because air expands more than water. As the air heats, it expands, pushing the water upward in the tube. One way to build such a thermometer is to put a straw in a soda can and to seal the opening of the can with modeling clay so that water can escape only by moving upward, out of the straw. It is important that student thermometers are tightly sealed. A sample design is shown below.

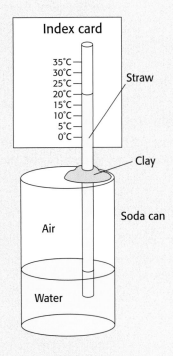

Answers

13. Answers will vary. Accept all reasonable responses.

14. Answers will vary. Accept all reasonable responses.

15. Accept all reasonable responses.

 Science Skills Worksheet "Being Flexible"

 Science Skills Worksheet "Using Logic"

Go Fly a Bike!
Teacher's Notes

Time Required

One 45-minute class period

Lab Ratings

EASY ——————→ HARD

TEACHER PREP	🥼
STUDENT SET-UP	🥼🥼🥼
CONCEPT LEVEL	🥼🥼
CLEAN UP	🥼🥼

MATERIALS

The materials listed on the student page are enough for a group of 3–4 students.

Safety Caution

Remind students to review all safety cautions and icons before beginning this lab activity.

Preparation Notes

Conduct this activity on a day when the wind is blowing, but not when the wind speed is greater than 50 km/h. Use straight, plastic straws. Before the activity, explain that an *anemometer* is a device that measures wind speed. It works because the wind pushes the cups at the same speed that the wind is moving.

 Datasheets for LabBook

Go Fly a Bike!

Your friend Daniel just invented a bicycle that can fly! Trouble is, the bike can fly only when the wind speed is between 3 m/s and 10 m/s. If the wind is not blowing hard enough, the bike won't get enough lift to rise into the air, and if the wind is blowing too hard, the bike is difficult to control. Daniel needs to know if he can fly his bike today. Can you build a device that can estimate how fast the wind is blowing?

Ask a Question

1. How can I construct a device to measure wind speed?

Construct an Anemometer

2. Cut off the rolled edges of all five paper cups. This will make them lighter, so that they can spin more easily.

3. Measure and place four equally spaced markings 1 cm below the rim of one of the paper cups.

4. Use the hole punch to punch a hole at each mark so that the cup has four equally spaced holes. Use the sharp pencil to carefully punch a hole in the center of the bottom of the cup.

5. Push a straw through two opposite holes in the side of the cup.

6. Repeat step 5 for the other two holes. The straws should form an X.

7. Measure 3 cm from the bottom of the remaining paper cups, and mark each spot with a dot.

8. At each dot, punch a hole in the paper cups with the hole punch.

9. Color the outside of one of the four cups.

10. Slide a cup on one of the straws by pushing the straw through the punched hole. Rotate the cup so that the bottom faces to the right.

Materials

- scissors
- 5 small paper cups
- metric ruler
- hole punch
- 2 straight plastic straws
- colored marker
- small stapler
- thumbtack
- sharp pencil with an eraser
- modeling clay
- masking tape
- watch or clock that indicates seconds

Terry J. Rakes
Elmwood Jr. High
Rogers, Arkansas

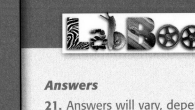

11. Fold the end of the straw, and staple it to the inside of the cup directly across from the hole.

12. Repeat steps 10–11 for each of the remaining cups.

13. Push the tack through the intersection of the two straws.

14. Push the eraser end of a pencil through the bottom hole in the center cup. Push the tack as far as it will go into the end of the eraser.

15. Push the sharpened end of the pencil into some modeling clay to form a base. This will allow the device to stand up without being knocked over, as shown at right.

16. Blow into the cups so that they spin. Adjust the tack so that the cups can freely spin without wobbling or falling apart. Congratulations! You have just constructed an anemometer.

Conduct an Experiment

17. Find a suitable area outside to place the anemometer vertically on a surface away from objects that would obstruct the wind, such as buildings and trees.

18. Mark the surface at the base of the anemometer with masking tape. Label the tape "starting point."

19. Hold the colored cup over the starting point while your partner holds the watch.

20. Release the colored cup. At the same time, your partner should look at the watch or clock. As the cups spin, count the number of times the colored cup crosses the starting point in 10 seconds.

Analyze the Results

21. How many times did the colored cup cross the starting point in 10 seconds?

22. Divide your answer in step 21 by 10 to get the number of revolutions in 1 second.

23. Measure the diameter of your anemometer (the distance between the outside edges of two opposite cups) in centimeters. Multiply this number by 3.14 to get the circumference of the circle made by the cups of your anemometer.

24. Multiply your answer from step 23 by the number of revolutions per second (step 22). Divide that answer by 100 to get wind speed in meters per second.

25. Compare your results with those of your classmates. Did you get the same result? What could account for any slight differences in your results?

Draw Conclusions

26. Could Daniel fly his bicycle today? Why or why not?

Answers

21. Answers will vary, depending on the wind speed.

22. Answers will vary according to each student's response to question 21.

23. Answers will vary according to the length of the straws and the size of the cups used.

24. Answers will vary.

25. Each group's anemometer should provide similar results. Differences may be caused by inconsistent wind speed. If students did not answer question 21 accurately, their results will be slightly different from other groups.

26. If the wind speed is between 3 and 10 m/s, Daniel could fly his bicycle. Otherwise, the weather would be too windy or too still for the bicycle to work.

103

Watching the Weather
Teacher's Notes

Time Required

One 45-minute class period

Lab Ratings

EASY ━━━━━━━━━━► HARD

TEACHER PREP ▲
STUDENT SET-UP ▲
CONCEPT LEVEL ▲▲
CLEAN UP ▲

MATERIALS

There are no materials required in this lab. Have students complete the lab individually.

Gordon Zibelman
Drexel Hill Middle School
Drexel Hill, Pennsylvania

Watching the Weather

Imagine that you own a private consulting firm that helps people plan for big occasions, such as weddings, parties, and celebrity events. One of your duties is making sure the weather doesn't put a damper on your clients' plans. In order to provide the best service possible, you have taken a crash course in reading weather maps. Will the celebrity golf match have to be delayed on account of rain? Will the wedding ceremony have to be moved inside so the blushing bride doesn't get soaked? It is your job to say "yea" or "nay."

Procedure TRY at HOME

1. Study the station model and legend shown on the next page. You will use the legend to interpret the weather map on the final page of this activity.

2. Weather data is represented on a weather map by a station model. A station model is a small circle that shows the location of the weather station along with a set of symbols and numbers around the circle that represent the data collected at the weather station. Study the table below.

Weather-Map Symbols					
Weather conditions		**Cloud cover**		**Wind speed (mph)**	
••	Light rain	○	No clouds	◎	Calm
∴	Moderate rain	◐	One-tenth or less		3–8
⁘	Heavy rain	◕	Two- to three-tenths		9–14
,	Drizzle	◑	Broken		15–20
＊＊	Light snow	◔	Nine-tenths		21–25
＊＊＊	Moderate snow	●	Overcast		32–37
R	Thunderstorm	⊗	Sky obscured		44–48
∽	Freezing rain		**Special Symbols**		55–60
∞	Haze	▲▲▲▲	Cold front		66–71
≡	Fog	●●●●	Warm front		
		H	High pressure		
		L	Low pressure		
		↺	Hurricane		

104

Station Model

Wind speed is represented by whole and half tails.

A line indicates the direction the wind is coming from.

Air temperature

A symbol represents the current weather conditions. If there is no symbol, there is no precipitation.

Dew point temperature

Shading indicates the cloud coverage.

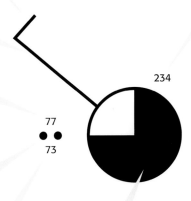

234

77

73

Atmospheric pressure in millibars (mbar). This number has been shortened on the station model. To read the number properly you must follow a few simple rules.
- If the first number is greater than 5, place a 9 in front of the number and a decimal point between the last two digits.
- If the first number is less than or equal to 5, place a 10 in front of the number and a decimal point between the last two digits.

Interpreting Station Models

The station model below is for Boston, Massachusetts. The current temperature in Boston is 42°F, and the dew point is 39°F. The barometric pressure is 1011.0 mbar. The sky is overcast, and there is a moderate rainfall. The wind is coming from the southwest at 15–20 mph.

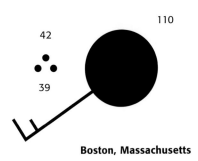

110

42

39

Boston, Massachusetts

Lab Notes

You may want to go over the different weather symbols with students and discuss how to convert the abbreviated form of atmospheric pressure to its actual measure. Before the lab, have students review the different kinds of fronts. Students may enjoy creating a weather report based on the weather report provided in this lab. Students can present this report to the class as a "live" studio show or through a video tape they create in their own time.

Datasheets for LabBook

105

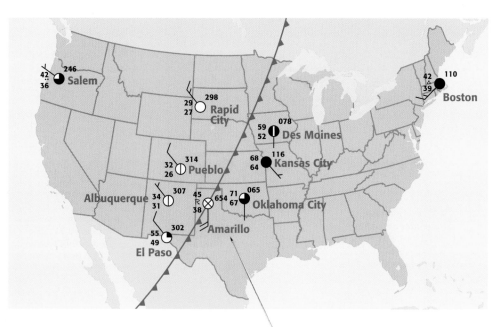

Answers

3. It's the winter. A cold front is coming through. Temperatures are low where the cold front has passed.

4. The temperature is 42°F. The dewpoint is 36°F. There is broken cloud cover, light rain, and the wind is from the northwest at 3–8 mph. The barometric pressure is 1024.6 mb.

5. As the cold front approaches, the wind is generally from the south, the barometric pressure is low, and temperatures are warmer. As the cold front passes, the wind is from the northwest, the pressure rises, and temperatures are much cooler.

6. The temperature is 45°F. The barometric pressure is 965.4 mb, the dewpoint is 38°F, the sky is obscured, there is a thunderstorm, and the wind is from the south at 21–25 mph.

Analysis

3. Based on the weather for the entire United States, what time of year is it? Explain your answer.

4. Interpret the station model for Salem, Oregon. What is the temperature, dew point, cloud coverage, wind direction, wind speed, and atmospheric pressure? Is there any precipitation? If so, what kind?

5. What is happening to wind direction, temperature, and pressure as the cold front approaches? as it passes?

6. Interpret the station model for Amarillo, Texas.

Let It Snow!

While an inch of rain might be good for your garden, 7 or 8 cm could cause an unwelcome flood. But what about snow? How much snow is too much? A blizzard might drop 40 cm of snow overnight. Sure it's up to your knees, but how does this much snow compare with rain? This activity will help you find out.

Procedure

1. Pour 50 mL of shaved ice into your beaker. Do not pack the ice into the beaker. This ice will represent your snowfall.

2. Use the ruler to measure the height of the snow in the beaker.

3. Turn on the hot plate to a low setting.
   Caution: Wear heat-resistant gloves and goggles when working with the hot plate.

4. Place the beaker on the hot plate, and leave it there until all of the snow melts.

5. Pour the water into the graduated cylinder, and record the height and volume of the water in your ScienceLog.

6. Repeat steps 1–5 two more times.

Analysis

7. What was the difference in height before and after the snow melted in each of your three trials? What was the average difference?

8. Why did the volume change after the ice melted?

9. In this activity, what was the ratio of snow height to water height?

10. Use the ratio you found in step 9 to calculate how much water 50 cm of this snow would produce. Use the following equation to help.

$$\frac{\text{measured height of snow}}{\text{measured height of water}} = \frac{50 \text{ cm of snow}}{? \text{ cm of water}}$$

11. Why is it important to know the water content of a snowfall?

Materials

- 150 mL of shaved ice
- 100 mL beaker
- metric ruler
- heat-resistant gloves
- hot plate
- graduated cylinder

Going Further

Shaved ice isn't really snow. Research to find out how much water real snow would produce. Does every snowfall produce the same ratio of snow height to water depth?

Walter Woolbaugh
Manhattan School System
Manhattan, Montana

Going Further

Every snowfall does not produce the same ratio of snow height to water depth. The ratio of snow height to water depth is dependent on several variables, including whether the snow is wet or dry.

Let It Snow!
Teacher's Notes

Time Required

One 45-minute class period

Lab Ratings

EASY	HARD

TEACHER PREP
STUDENT SET-UP
CONCEPT LEVEL
CLEAN UP

MATERIALS

The materials listed are best for a group of 3–4 students.

Safety Caution

Remind students to review all safety cautions and icons before beginning this lab activity.

Answers

7. Answers will vary according to the water content of the ice or snow sample.

8. The volume changed because the water changed from a solid to a liquid.

9. Answers will vary.

10. Answers will vary

11. Answers will vary. The water content of a snowfall—whether it is relatively wet or relatively dry—affects how much flooding may occur as the snow melts. A "wetter" snow has more water per volume and may cause more flooding than a "drier" snow.

 Datasheets for LabBook

Global Impact
Teacher's Notes

Time Required
One 45-minute class period

Lab Ratings

EASY ——————————→ HARD

TEACHER PREP 🧪
STUDENT SET-UP 🧪🧪🧪
CONCEPT LEVEL 🧪🧪
CLEAN UP 🧪

MATERIALS
The materials listed in the student page are enough for each student.

Preparation Notes

This activity requires graphing skills. Students may need a review of graphing, analyzing data from a graph, and calculating the slope of a graph.

Datasheets for LabBook

Science Skills Worksheet
"Grasping Graphing"

Janel Guse
West Central Middle School
Hartford, South Dakota

Global Impact

SKILL BUILDER

For years scientists have debated the topic of global warming. Is the temperature of the Earth actually getting warmer? Sample sizes are a very important factor in any scientific study. In this activity, you will examine a chart to determine if the data indicate any trends. Be sure to notice how much the trends seem to change as you analyze different sets of data.

Materials
- 4 colored pencils
- metric ruler

TRY at HOME

Procedure

1. Look at the chart below. It shows average global temperatures recorded over the last 100 years.

2. Draw a graph in your ScienceLog. Label the horizontal axis "Time," and mark the grid in 5-year intervals. Label the vertical axis "Temperature (°C)," with values ranging from 13°C to 15°C.

3. Starting with 1900, use the numbers in red to plot the temperature in 20-year intervals. Connect the dots with straight lines.

4. Using a ruler, estimate the overall slope of temperatures, and draw a red line to represent the slope.

5. Using different colors, plot the temperatures at 10-year intervals and 5-year intervals on the same graph. Connect each set of dots, and draw the average slope for each set.

Analysis

6. Examine your completed graph, and explain any trends you see in the graphed data. Was there an increase or a decrease in average temperature over the last 100 years?

7. What differences did you see in each set of graphed data? what similarities?

8. What conclusions can you draw from the data you graphed in this activity?

9. What would happen if your graph were plotted in 1-year intervals? Try it!

Average Global Temperatures											
Year	°C	Year	°C	Year	°C	Year	°C	Year	°C	Year	°C
1900	14.0	1917	13.6	1934	14.0	1951	14.0	1968	13.9	1985	14.1
1901	13.9	1918	13.6	1935	13.9	1952	14.0	1969	14.0	1986	14.2
1902	13.8	1919	13.8	1936	14.0	1953	14.1	1970	14.0	1987	14.3
1903	13.6	1920	13.8	1937	14.1	1954	13.9	1971	13.9	1988	14.4
1904	13.5	1921	13.9	1938	14.1	1955	13.9	1972	13.9	1989	14.2
1905	13.7	1922	13.9	1939	14.0	1956	13.8	1973	14.2	1990	14.5
1906	13.8	1923	13.8	1940	14.1	1957	14.1	1974	13.9	1991	14.4
1907	13.6	1924	13.8	1941	14.1	1958	14.1	1975	14.0	1992	14.1
1908	13.7	1925	13.8	1942	14.1	1959	14.0	1976	13.8	1993	14.2
1909	13.7	1926	14.1	1943	14.0	1960	14.0	1977	14.2	1994	14.3
1910	13.7	1927	14.0	1944	14.1	1961	14.1	1978	14.1	1995	14.5
1911	13.7	1928	14.0	1945	14.0	1962	14.0	1979	14.1	1996	14.4
1912	13.7	1929	13.8	1946	14.0	1963	14.0	1980	14.3	1997	14.4
1913	13.8	1930	13.9	1947	14.1	1964	13.7	1981	14.4	1998	14.5
1914	14.0	1931	14.0	1948	14.0	1965	13.8	1982	14.1	1999	
1915	14.0	1932	14.0	1949	13.9	1966	13.9	1983	14.3	2000	
1916	13.8	1933	13.9	1950	13.8	1967	14.0	1984	14.1	2001	

108

Answers

6. Students will notice that the temperatures fluctuated over the last 100 years but have gradually increased in the last 30 years.

7. The larger the sample size, the more precise your analysis will be. You notice that in the larger samples, temperatures are constantly fluctuating. A smaller sample doesn't always adequately display what is really happening with the data. For instance, the average temperature for a certain year might not be representative for the entire decade. The smaller the data set, the more your outcome is subject to error. There were very few similarities among the graphs.

8. You can conclude that a larger data set gives you a more complete picture of what is happening. Global temperatures have gradually increased in the twentieth century.

9. Global temperatures would appear to fluctuate more.

For the Birds

You and a partner have a new business building birdhouses. But your first clients have told you that birds do not want to live in the birdhouses you have made. The clients want their money back unless you can solve the problem. You need to come up with a solution right away!

You remember reading an article about microclimates in a science magazine. Cities often heat up because the pavement and buildings absorb so much solar radiation. Maybe the houses are too warm! How can the houses be kept cooler?

You decide to investigate the roofs; after all, changing the roofs would be a lot easier than building new houses. In order to help your clients and the birds, you decide to test different roof colors and materials to see how these variables affect a roof's ability to absorb the sun's rays.

One partner will test the color, and the other partner will test the materials. You will then share your results and make a recommendation together.

Materials

- 4 pieces of cardboard
- black, white, and light-blue tempera paint
- 4 Celsius thermometers
- watch or clock
- beige or tan wood
- beige or tan rubber

Part A: Color Test

Ask a Question

1. What color would be the best choice for the roof of a birdhouse?

Form a Hypothesis

2. In your ScienceLog, write down the color you think will keep a birdhouse coolest.

Test the Hypothesis

3. Paint one piece of cardboard black, another piece white, and a third light blue.

4. After the paint has dried, take the three pieces of cardboard outside, and place a thermometer on each piece.

5. In an area where there is no shade, place each piece at the same height so that all three receive the same amount of sunlight. Leave the pieces in the sunlight for 15 minutes.

6. Leave a fourth thermometer outside in the shade to measure the temperature of the air.

For the Birds
Teacher's Notes

Time Required

One 45-minute class period

Lab Ratings

EASY ——————————→ HARD

TEACHER PREP	🧪
STUDENT SET-UP	🧪🧪
CONCEPT LEVEL	🧪🧪
CLEAN UP	🧪🧪

MATERIALS

The materials listed on the student page are enough for a group of 4–5 students.

Safety Caution

Remind students to review all safety cautions and icons before beginning this lab activity.

 Datasheets for LabBook

Larry Tackett
Andrew Jackson Middle School
Cross Lanes, West Virginia

Answers

8. Sample answer: No, the thermometers recorded different temperatures. The black and blue pieces of cardboard, particularly the black one, caused the temperature to increase.

9. Sample answer: The temperature of the black cardboard was much higher than the outside temperature. Students should find that the temperature of the other colors was also different from the outside temperature.

10. Students' answers will vary. Accept all reasonable responses.

7. In your ScienceLog, record the reading of the thermometer on each piece of cardboard. Also record the outside temperature.

Analyze the Results

8. Did each of the three thermometers record the same temperature after 15 minutes? Explain.

Part B: Material Test

Ask a Question

11. Which material would be the best choice for the roof of a birdhouse?

Form a Hypothesis

12. In your ScienceLog, write down the material you think will keep a birdhouse coolest.

Test the Hypothesis

13. Take the rubber, wood, and the fourth piece of cardboard outside, and place a thermometer on each.

9. Were the temperature readings on each of the three pieces of cardboard the same as the reading for the outside temperature? Explain.

Draw Conclusions

10. How do your observations compare with your hypothesis?

14. In an area where there is no shade, place each material at the same height so that they all receive the same amount of sunlight. Leave the materials in the sunlight for 15 minutes.

15. Leave a fourth thermometer outside in the shade to measure the temperature of the air.

16. In your ScienceLog, record the temperature of each material. Also record the outside temperature.

110

Analyze the Results

17. Did each of the thermometers on the three materials record the same temperature after 15 minutes? Explain.

18. Were the temperature readings on the rubber, wood, and cardboard the same as the reading for the outside temperature? Explain.

Sharing Information (Parts A and B)

Communicate Results

After you and your partner have finished your investigations, take a few minutes to share your results. Then work together to design a new roof.

20. Which material would you use to build the roofs for your birdhouses? Why?

21. Which color would you use to paint the new roofs? Why?

Draw Conclusions

19. How do your observations compare with your hypothesis?

Going Further

Make three different-colored samples for each of the three materials. When you measure the temperatures for each sample, how do the colors compare for each material? Is the same color best for all three materials? How do your results compare with what you concluded in steps 20 and 21 of this activity? What's more important, color or material?

Answers

17. Sample answer: No, the temperatures were different. The temperature of the rubber was higher than that of the other two materials.

18. Sample answer: No, the temperature of the rubber was higher than the outside temperature. Accept all other reasonable answers for the other materials.

19. Answers will vary. Accept all reasonable answers.

20. Sample answer: The wood would be the coolest. The cardboard would be a possible alternative.

21. The white roof would be the coolest. A light blue roof would be a possible alternative.

Going Further

Answers will vary. Accept all reasonable interpretations of the data collected.

Science Skills Worksheet "Understanding Variables"

Concept Mapping: A Way to Bring Ideas Together

What Is a Concept Map?

Have you ever tried to tell someone about a book or a chapter you've just read and found that you can remember only a few isolated words and ideas? Or maybe you've memorized facts for a test and then weeks later discovered you're not even sure what topics those facts covered.

In both cases, you may have understood the ideas or concepts by themselves but not in relation to one another. If you could somehow link the ideas together, you would probably understand them better and remember them longer. This is something a concept map can help you do. A concept map is a way to see how ideas or concepts fit together. It can help you see the "big picture."

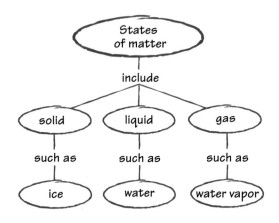

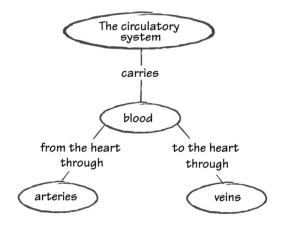

How to Make a Concept Map

1 **Make a list of the main ideas or concepts.**

It might help to write each concept on its own slip of paper. This will make it easier to rearrange the concepts as many times as necessary to make sense of how the concepts are connected. After you've made a few concept maps this way, you can go directly from writing your list to actually making the map.

2 **Arrange the concepts in order from the most general to the most specific.**

Put the most general concept at the top and circle it. Ask yourself, "How does this concept relate to the remaining concepts?" As you see the relationships, arrange the concepts in order from general to specific.

3 **Connect the related concepts with lines.**

4 **On each line, write an action word or short phrase that shows how the concepts are related.**

Look at the concept maps on this page, and then see if you can make one for the following terms:

plants, water, photosynthesis, carbon dioxide, sun's energy

One possible answer is provided at right, but don't look at it until you try the concept map yourself.

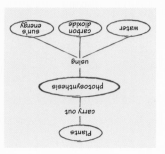

SI Measurement

The International System of Units, or SI, is the standard system of measurement used by many scientists. Using the same standards of measurement makes it easier for scientists to communicate with one another.

SI works by combining prefixes and base units. Each base unit can be used with different prefixes to define smaller and larger quantities. The table below lists common SI prefixes.

SI Prefixes			
Prefix	**Abbreviation**	**Factor**	**Example**
kilo-	k	1,000	kilogram, 1 kg = 1,000 g
hecto-	h	100	hectoliter, 1 hL = 100 L
deka-	da	10	dekameter, 1 dam = 10 m
		1	meter, liter
deci-	d	0.1	decigram, 1 dg = 0.1 g
centi-	c	0.01	centimeter, 1 cm = 0.01 m
milli-	m	0.001	milliliter, 1 mL = 0.001 L
micro-	μ	0.000 001	micrometer, 1 μm = 0.000 001 m

SI Conversion Table		
SI units	**From SI to English**	**From English to SI**
Length		
kilometer (km) = 1,000 m	1 km = 0.621 mi	1 mi = 1.609 km
meter (m) = 100 cm	1 m = 3.281 ft	1 ft = 0.305 m
centimeter (cm) = 0.01 m	1 cm = 0.394 in.	1 in. = 2.540 cm
millimeter (mm) = 0.001 m	1 mm = 0.039 in.	
micrometer (μm) = 0.000 001 m		
nanometer (nm) = 0.000 000 001 m		
Area		
square kilometer (km^2) = 100 hectares	1 km^2 = 0.386 mi^2	1 mi^2 = 2.590 km^2
hectare (ha) = 10,000 m^2	1 ha = 2.471 acres	1 acre = 0.405 ha
square meter (m^2) = 10,000 cm^2	1 m^2 = 10.765 ft^2	1 ft^2 = 0.093 m^2
square centimeter (cm^2) = 100 mm^2	1 cm^2 = 0.155 $in.^2$	1 $in.^2$ = 6.452 cm^2
Volume		
liter (L) = 1,000 mL = 1 dm^3	1 L = 1.057 fl qt	1 fl qt = 0.946 L
milliliter (mL) = 0.001 L = 1 cm^3	1 mL = 0.034 fl oz	1 fl oz = 29.575 mL
microliter (μL) = 0.000 001 L		
Mass		
kilogram (kg) = 1,000 g	1 kg = 2.205 lb	1 lb = 0.454 kg
gram (g) = 1,000 mg	1 g = 0.035 oz	1 oz = 28.349 g
milligram (mg) = 0.001 g		
microgram (μg) = 0.000 001 g		

Temperature Scales

Temperature can be expressed using three different scales: Fahrenheit, Celsius, and Kelvin. The SI unit for temperature is the kelvin (K).

Although 0 K is much colder than 0°C, a change of 1 K is equal to a change of 1°C.

Three Temperature Scales

	Fahrenheit	Celsius	Kelvin
Water boils	212°	100°	373
Body temperature	98.6°	37°	310
Room temperature	68°	20°	293
Water freezes	32°	0°	273

Temperature Conversions Table

To convert	Use this equation:	Example
Celsius to Fahrenheit °C ⟶ °F	$°F = \left(\dfrac{9}{5} \times °C\right) + 32$	Convert 45°C to °F. $°F = \left(\dfrac{9}{5} \times 45°C\right) + 32 = 113°F$
Fahrenheit to Celsius °F ⟶ °C	$°C = \dfrac{5}{9} \times (°F - 32)$	Convert 68°F to °C. $°C = \dfrac{5}{9} \times (68°F - 32) = 20°C$
Celsius to Kelvin °C ⟶ K	$K = °C + 273$	Convert 45°C to K. $K = 45°C + 273 = 318\ K$
Kelvin to Celsius K ⟶ °C	$°C = K - 273$	Convert 32 K to °C. $°C = 32\ K - 273 = -241°C$

114

114   Appendix

Measuring Skills

Using a Graduated Cylinder

When using a graduated cylinder to measure volume, keep the following procedures in mind:

1 Make sure the cylinder is on a flat, level surface.

2 Move your head so that your eye is level with the surface of the liquid.

3 Read the mark closest to the liquid level. On glass graduated cylinders, read the mark closest to the center of the curve in the liquid's surface.

Using a Meterstick or Metric Ruler

When using a meterstick or metric ruler to measure length, keep the following procedures in mind:

1 Place the ruler firmly against the object you are measuring.

2 Align one edge of the object exactly with the zero end of the ruler.

3 Look at the other edge of the object to see which of the marks on the ruler is closest to that edge. **Note:** Each small slash between the centimeters represents a millimeter, which is one-tenth of a centimeter.

Using a Triple-Beam Balance

When using a triple-beam balance to measure mass, keep the following procedures in mind:

1 Make sure the balance is on a level surface.

2 Place all of the countermasses at zero. Adjust the balancing knob until the pointer rests at zero.

3 Place the object you wish to measure on the pan. **Caution:** Do not place hot objects or chemicals directly on the balance pan.

4 Move the largest countermass along the beam to the right until it is at the last notch that does not tip the balance. Follow the same procedure with the next-largest countermass. Then move the smallest countermass until the pointer rests at zero.

5 Add the readings from the three beams together to determine the mass of the object.

6 When determining the mass of crystals or powders, use a piece of filter paper. First find the mass of the paper. Then add the crystals or powder to the paper and re-measure. The actual mass of the crystals or powder is the total mass minus the mass of the paper. When finding the mass of liquids, first find the mass of the empty container. Then find the mass of the liquid and container together. The mass of the liquid is the total mass minus the mass of the container.

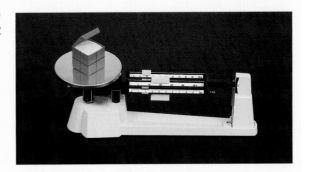

Scientific Method

The series of steps that scientists use to answer questions and solve problems is often called the **scientific method.** The scientific method is not a rigid procedure. Scientists may use all of the steps or just some of the steps of the scientific method. They may even repeat some of the steps. The goal of the scientific method is to come up with reliable answers and solutions.

Six Steps of the Scientific Method

Ask a Question

1 **Ask a Question** Good questions come from careful **observations.** You make observations by using your senses to gather information. Sometimes you may use instruments, such as microscopes and telescopes, to extend the range of your senses. As you observe the natural world, you will discover that you have many more questions than answers. These questions drive the scientific method.

Questions beginning with *what, why, how,* and *when* are very important in focusing an investigation, and they often lead to a hypothesis. (You will learn what a hypothesis is in the next step.) Here is an example of a question that could lead to further investigation.

Question: How does acid rain affect plant growth?

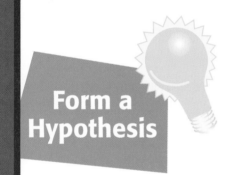

Form a Hypothesis

2 **Form a Hypothesis** After you come up with a question, you need to turn the question into a **hypothesis.** A hypothesis is a clear statement of what you expect the answer to your question to be. Your hypothesis will represent your best "educated guess" based on your observations and what you already know. A good hypothesis is testable. If observations and information cannot be gathered or if an experiment cannot be designed to test your hypothesis, it is untestable, and the investigation can go no further.

Here is a hypothesis that could be formed from the question, "How does acid rain affect plant growth?"

Hypothesis: Acid rain causes plants to grow more slowly.

Notice that the hypothesis provides some specifics that lead to methods of testing. The hypothesis can also lead to predictions. A **prediction** is what you think will be the outcome of your experiment or data collection. Predictions are usually stated in an "if . . . then" format. For example, **if** meat is kept at room temperature, **then** it will spoil faster than meat kept in the refrigerator. More than one prediction can be made for a single hypothesis. Here is a sample prediction for the hypothesis that acid rain causes plants to grow more slowly.

Prediction: If a plant is watered with only acid rain (which has a pH of 4), then the plant will grow at half its normal rate.

3 **Test the Hypothesis** After you have formed a hypothesis and made a prediction, you should test your hypothesis. There are different ways to do this. Perhaps the most familiar way is to conduct a **controlled experiment.** A controlled experiment tests only one factor at a time. A controlled experiment has a **control group** and one or more **experimental groups.** All the factors for the control and experimental groups are the same except for one factor, which is called the **variable.** By changing only one factor, you can see the results of just that one change.

Sometimes, the nature of an investigation makes a controlled experiment impossible. For example, dinosaurs have been extinct for millions of years, and the Earth's core is surrounded by thousands of meters of rock. It would be difficult, if not impossible, to conduct controlled experiments on such things. Under such circumstances, a hypothesis may be tested by making detailed observations. Taking measurements is one way of making observations.

Test the Hypothesis

4 **Analyze the Results** After you have completed your experiments, made your observations, and collected your data, you must analyze all the information you have gathered. Tables and graphs are often used in this step to organize the data.

Analyze the Results

5 **Draw Conclusions** Based on the analysis of your data, you should conclude whether or not your results support your hypothesis. If your hypothesis is supported, you (or others) might want to repeat the observations or experiments to verify your results. If your hypothesis is not supported by the data, you may have to check your procedure for errors. You may even have to reject your hypothesis and make a new one. If you cannot draw a conclusion from your results, you may have to try the investigation again or carry out further observations or experiments.

Draw Conclusions

Do they support your hypothesis?

No

Yes

6 **Communicate Results** After any scientific investigation, you should report your results. By doing a written or oral report, you let others know what you have learned. They may want to repeat your investigation to see if they get the same results. Your report may even lead to another question, which in turn may lead to another investigation.

Communicate Results

Scientific Method in Action

The scientific method is not a "straight line" of steps. It contains loops in which several steps may be repeated over and over again, while others may not be necessary. For example, sometimes scientists will find that testing one hypothesis raises new questions and new hypotheses to be tested. And sometimes, testing the hypothesis leads directly to a conclusion. Furthermore, the steps in the scientific method are not always used in the same order. Follow the steps in the diagram below, and see how many different directions the scientific method can take you.

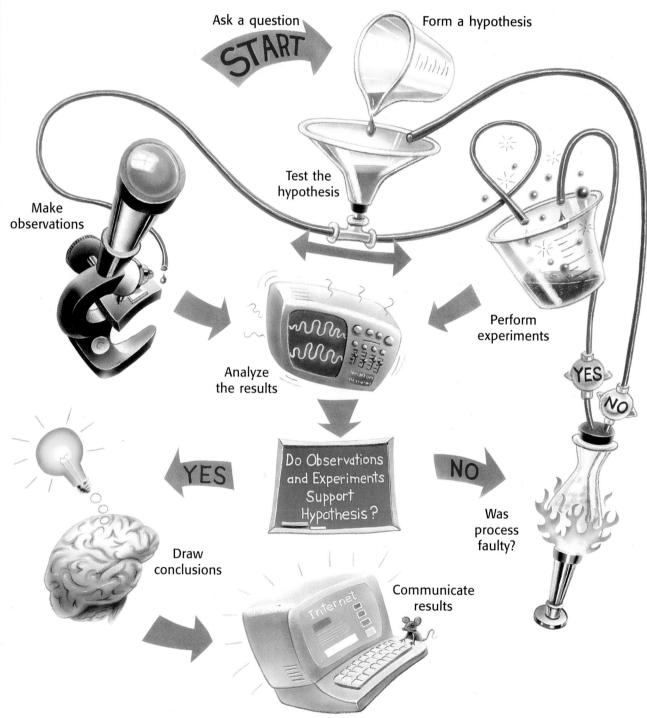

Ask a question

START

Form a hypothesis

Test the hypothesis

Make observations

Perform experiments

Analyze the results

YES

NO

YES

Do Observations and Experiments Support Hypothesis?

NO

Draw conclusions

Was process faulty?

Communicate results

Internet

Making Charts and Graphs

Circle Graphs

A circle graph, or pie chart, shows how each group of data relates to all of the data. Each part of the circle represents a category of the data. The entire circle represents all of the data. For example, a biologist studying a hardwood forest in Wisconsin found that there were five different types of trees. The data table at right summarizes the biologist's findings.

Wisconsin Hardwood Trees	
Type of tree	**Number found**
Oak	600
Maple	750
Beech	300
Birch	1,200
Hickory	150
Total	3,000

How to Make a Circle Graph

1 In order to make a circle graph of this data, first find the percentage of each type of tree. To do this, divide the number of individual trees by the total number of trees and multiply by 100.

$$\frac{600 \text{ oak}}{3,000 \text{ trees}} \times 100 = 20\%$$

$$\frac{750 \text{ maple}}{3,000 \text{ trees}} \times 100 = 25\%$$

$$\frac{300 \text{ beech}}{3,000 \text{ trees}} \times 100 = 10\%$$

$$\frac{1,200 \text{ birch}}{3,000 \text{ trees}} \times 100 = 40\%$$

$$\frac{150 \text{ hickory}}{3,000 \text{ trees}} \times 100 = 5\%$$

2 Now determine the size of the pie shapes that make up the chart. Do this by multiplying each percentage by 360°. Remember that a circle contains 360°.

$20\% \times 360° = 72°$ $25\% \times 360° = 90°$
$10\% \times 360° = 36°$ $40\% \times 360° = 144°$
$5\% \times 360° = 18°$

3 Then check that the sum of the percentages is 100 and the sum of the degrees is 360.

$20\% + 25\% + 10\% + 40\% + 5\% = 100\%$
$72° + 90° + 36° + 144° + 18° = 360°$

4 Use a compass to draw a circle and mark its center.

5 Then use a protractor to draw angles of 72°, 90°, 36°, 144°, and 18° in the circle.

6 Finally, label each part of the graph, and choose an appropriate title.

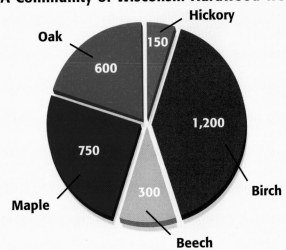

A Community of Wisconsin Hardwood Trees

Line Graphs

Population of Appleton, 1900–2000	
Year	Population
1900	1,800
1920	2,500
1940	3,200
1960	3,900
1980	4,600
2000	5,300

Line graphs are most often used to demonstrate continuous change. For example, Mr. Smith's science class analyzed the population records for their hometown, Appleton, between 1900 and 2000. Examine the data at left.

Because the year and the population change, they are the *variables*. The population is determined by, or dependent on, the year. Therefore, the population is called the **dependent variable**, and the year is called the **independent variable**. Each set of data is called a **data pair**. To prepare a line graph, data pairs must first be organized in a table like the one at left.

How to Make a Line Graph

❶ Place the independent variable along the horizontal (*x*) axis. Place the dependent variable along the vertical (*y*) axis.

❷ Label the *x*-axis "Year" and the *y*-axis "Population." Look at your largest and smallest values for the population. Determine a scale for the *y*-axis that will provide enough space to show these values. You must use the same scale for the entire length of the axis. Find an appropriate scale for the *x*-axis too.

❸ Choose reasonable starting points for each axis.

❹ Plot the data pairs as accurately as possible.

❺ Choose a title that accurately represents the data.

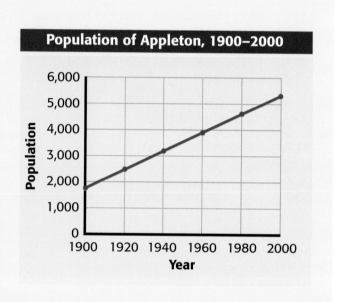

Population of Appleton, 1900–2000

How to Determine Slope

Slope is the ratio of the change in the *y*-axis to the change in the *x*-axis, or "rise over run."

❶ Choose two points on the line graph. For example, the population of Appleton in 2000 was 5,300 people. Therefore, you can define point *a* as (2000, 5,300). In 1900, the population was 1,800 people. Define point *b* as (1900, 1,800).

❷ Find the change in the *y*-axis.
(*y* at point *a*) − (*y* at point *b*)
5,300 people − 1,800 people = 3,500 people

❸ Find the change in the *x*-axis.
(*x* at point *a*) − (*x* at point *b*)
2000 − 1900 = 100 years

❹ Calculate the slope of the graph by dividing the change in *y* by the change in *x*.

$$\text{slope} = \frac{\text{change in } y}{\text{change in } x}$$

$$\text{slope} = \frac{3{,}500 \text{ people}}{100 \text{ years}}$$

slope = 35 people per year

In this example, the population in Appleton increased by a fixed amount each year. The graph of this data is a straight line. Therefore, the relationship is **linear.** When the graph of a set of data is not a straight line, the relationship is **nonlinear.**

Using Algebra to Determine Slope

The equation in step 4 may also be arranged to be:

$$y = kx$$

where y represents the change in the y-axis, k represents the slope, and x represents the change in the x-axis.

$$\text{slope} = \frac{\text{change in } y}{\text{change in } x}$$

$$k = \frac{y}{x}$$

$$k \times x = \frac{y \times x}{x}$$

$$kx = y$$

Bar Graphs

Bar graphs are used to demonstrate change that is not continuous. These graphs can be used to indicate trends when the data are taken over a long period of time. A meteorologist gathered the precipitation records at right for Hartford, Connecticut, for April 1–15, 1996, and used a bar graph to represent the data.

Precipitation in Hartford, Connecticut April 1–15, 1996

Date	Precipitation (cm)	Date	Precipitation (cm)
April 1	0.5	April 9	0.25
April 2	1.25	April 10	0.0
April 3	0.0	April 11	1.0
April 4	0.0	April 12	0.0
April 5	0.0	April 13	0.25
April 6	0.0	April 14	0.0
April 7	0.0	April 15	6.50
April 8	1.75		

How to Make a Bar Graph

❶ Use an appropriate scale and a reasonable starting point for each axis.

❷ Label the axes, and plot the data.

❸ Choose a title that accurately represents the data.

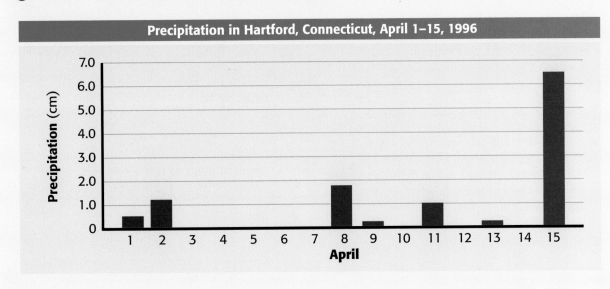

Precipitation in Hartford, Connecticut, April 1–15, 1996

Math Refresher

Science requires an understanding of many math concepts. The following pages will help you review some important math skills.

Averages

An **average,** or **mean,** simplifies a list of numbers into a single number that *approximates* their value.

> **Example:** Find the average of the following set of numbers: 5, 4, 7, and 8.

Step 1: Find the sum.

$$5 + 4 + 7 + 8 = 24$$

Step 2: Divide the sum by the amount of numbers in your set. Because there are four numbers in this example, divide the sum by 4.

$$\frac{24}{4} = 6$$

The average, or mean, is **6.**

Ratios

A **ratio** is a comparison between numbers, and it is usually written as a fraction.

> **Example:** Find the ratio of thermometers to students if you have 36 thermometers and 48 students in your class.

Step 1: Make the ratio.

$$\frac{36 \text{ thermometers}}{48 \text{ students}}$$

Step 2: Reduce the fraction to its simplest form.

$$\frac{36}{48} = \frac{36 \div 12}{48 \div 12} = \frac{3}{4}$$

The ratio of thermometers to students is **3 to 4,** or $\frac{3}{4}$. The ratio may also be written in the form 3:4.

Proportions

A **proportion** is an equation that states that two ratios are equal.

$$\frac{3}{1} = \frac{12}{4}$$

To solve a proportion, first multiply across the equal sign. This is called cross-multiplication. If you know three of the quantities in a proportion, you can use cross-multiplication to find the fourth.

> **Example:** Imagine that you are making a scale model of the solar system for your science project. The diameter of Jupiter is 11.2 times the diameter of the Earth. If you are using a plastic-foam ball with a diameter of 2 cm to represent the Earth, what diameter does the ball representing Jupiter need to be?
>
> $$\frac{11.2}{1} = \frac{x}{2 \text{ cm}}$$

Step 1: Cross-multiply.

$$\frac{11.2}{1} \diagdown\!\!\!\!\diagup \frac{x}{2}$$

$$11.2 \times 2 = x \times 1$$

Step 2: Multiply.

$$22.4 = x \times 1$$

Step 3: Isolate the variable by dividing both sides by 1.

$$x = \frac{22.4}{1}$$

$$x = 22.4 \text{ cm}$$

You will need to use a ball with a diameter of **22.4 cm** to represent Jupiter.

Percentages

A **percentage** is a ratio of a given number to 100.

> **Example:** What is 85 percent of 40?

Step 1: Rewrite the percentage by moving the decimal point two places to the left.

$$.85$$

Step 2: Multiply the decimal by the number you are calculating the percentage of.

$$0.85 \times 40 = 34$$

85 percent of 40 is **34.**

Decimals

To **add** or **subtract decimals,** line up the digits vertically so that the decimal points line up. Then add or subtract the columns from right to left, carrying or borrowing numbers as necessary.

> **Example:** Add the following numbers: 3.1415 and 2.96.

Step 1: Line up the digits vertically so that the decimal points line up.

$$\begin{array}{r} 3.1415 \\ + 2.96 \\ \hline \end{array}$$

Step 2: Add the columns from right to left, carrying when necessary.

$$\begin{array}{r} {}^{1\ 1} \\ 3.1415 \\ + 2.96 \\ \hline 6.1015 \end{array}$$

The sum is **6.1015.**

Fractions

Numbers tell you how many; **fractions** tell you *how much of a whole.*

> **Example:** Your class has 24 plants. Your teacher instructs you to put 5 in a shady spot. What fraction does this represent?

Step 1: Write a fraction with the total number of parts in the whole as the denominator.

$$\frac{?}{24}$$

Step 2: Write the number of parts of the whole being represented as the numerator.

$$\frac{5}{24}$$

$\frac{5}{24}$ of the plants will be in the shade.

Reducing Fractions

It is usually best to express a fraction in simplest form. This is called *reducing* a fraction.

> **Example:** Reduce the fraction $\frac{30}{45}$ to its simplest form.

Step 1: Find the largest whole number that will divide evenly into both the numerator and denominator. This number is called the greatest common factor (GCF).

factors of the numerator 30: 1, 2, 3, 5, 6, 10, **15,** 30

factors of the denominator 45: 1, 3, 5, 9, **15,** 45

Step 2: Divide both the numerator and the denominator by the GCF, which in this case is 15.

$$\frac{30}{45} = \frac{30 \div 15}{45 \div 15} = \frac{2}{3}$$

$\frac{30}{45}$ reduced to its simplest form is $\frac{2}{3}$.

Adding and Subtracting Fractions

To **add** or **subtract fractions** that have the **same denominator,** simply add or subtract the numerators.

Examples:
$$\frac{3}{5} + \frac{1}{5} = ? \quad \text{and} \quad \frac{3}{4} - \frac{1}{4} = ?$$

Step 1: Add or subtract the numerators.
$$\frac{3}{5} + \frac{1}{5} = \frac{4}{} \quad \text{and} \quad \frac{3}{4} - \frac{1}{4} = \frac{2}{}$$

Step 2: Write the sum or difference over the denominator.
$$\frac{3}{5} + \frac{1}{5} = \frac{4}{5} \quad \text{and} \quad \frac{3}{4} - \frac{1}{4} = \frac{2}{4}$$

Step 3: If necessary, reduce the fraction to its simplest form.
$$\frac{4}{5} \text{ cannot be reduced, and } \frac{2}{4} = \frac{1}{2}.$$

To **add** or **subtract fractions** that have **different denominators,** first find the least common denominator (LCD).

Examples:
$$\frac{1}{2} + \frac{1}{6} = ? \quad \text{and} \quad \frac{3}{4} - \frac{2}{3} = ?$$

Step 1: Write the equivalent fractions with a common demominator.
$$\frac{3}{6} + \frac{1}{6} = ? \quad \text{and} \quad \frac{9}{12} - \frac{8}{12} = ?$$

Step 2: Add or subtract.
$$\frac{3}{6} + \frac{1}{6} = \frac{4}{6} \quad \text{and} \quad \frac{9}{12} - \frac{8}{12} = \frac{1}{12}$$

Step 3: If necessary, reduce the fraction to its simplest form.
$$\frac{4}{6} = \frac{2}{3}, \text{ and } \frac{1}{12} \text{ cannot be reduced.}$$

Multiplying Fractions

To **multiply fractions,** multiply the numerators and the denominators together, and then reduce the fraction to its simplest form.

Example:
$$\frac{5}{9} \times \frac{7}{10} = ?$$

Step 1: Multiply the numerators and denominators.
$$\frac{5}{9} \times \frac{7}{10} = \frac{5 \times 7}{9 \times 10} = \frac{35}{90}$$

Step 2: Reduce.
$$\frac{35}{90} = \frac{35 \div 5}{90 \div 5} = \frac{7}{18}$$

Dividing Fractions

To **divide fractions,** first rewrite the divisor (the number you divide *by*) upside down. This is called the reciprocal of the divisor. Then you can multiply and reduce if necessary.

Example:
$$\frac{5}{8} \div \frac{3}{2} = ?$$

Step 1: Rewrite the divisor as its reciprocal.
$$\frac{3}{2} \rightarrow \frac{2}{3}$$

Step 2: Multiply.
$$\frac{5}{8} \times \frac{2}{3} = \frac{5 \times 2}{8 \times 3} = \frac{10}{24}$$

Step 3: Reduce.
$$\frac{10}{24} = \frac{10 \div 2}{24 \div 2} = \frac{5}{12}$$

Scientific Notation

Scientific notation is a short way of representing very large and very small numbers without writing all of the place-holding zeros.

> **Example:** Write 653,000,000 in scientific notation.

Step 1: Write the number without the place-holding zeros.

$$653$$

Step 2: Place the decimal point after the first digit.

$$6.53$$

Step 3: Find the exponent by counting the number of places that you moved the decimal point.

$$6.53000000$$

The decimal point was moved eight places to the left. Therefore, the exponent of 10 is positive 8. Remember, if the decimal point had moved to the right, the exponent would be negative.

Step 4: Write the number in scientific notation.

$$6.53 \times 10^8$$

Area

Area is the number of square units needed to cover the surface of an object.

> **Formulas:**
> Area of a square = side × side
> Area of a rectangle = length × width
> Area of a triangle = $\frac{1}{2}$ × base × height
>
> **Examples:** Find the areas.

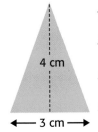

Triangle
Area = $\frac{1}{2}$ × base × height
Area = $\frac{1}{2}$ × 3 cm × 4 cm
Area = **6 cm²**

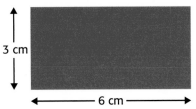

Rectangle
Area = length × width
Area = 6 cm × 3 cm
Area = **18 cm²**

Square
Area = side × side
Area = 3 cm × 3 cm
Area = **9 cm²**

Volume

Volume is the amount of space something occupies.

> **Formulas:**
> Volume of a cube =
> side × side × side
>
> Volume of a prism =
> area of base × height
>
> **Examples:**
> Find the volume
> of the solids.

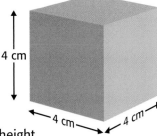

Cube
Volume = side × side × side
Volume = 4 cm × 4 cm × 4 cm
Volume = **64 cm³**

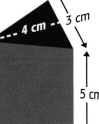

Prism
Volume = area of base × height
Volume = (area of triangle) × height
Volume = $\left(\frac{1}{2} \times 3 \text{ cm} \times 4 \text{ cm} \right) \times 5$ cm
Volume = 6 cm² × 5 cm
Volume = **30 cm³**

Physical Laws and Equations

Law of Conservation of Energy

The law of conservation of energy states that energy can be neither created nor destroyed.

The total amount of energy in a closed system is always the same. Energy can be changed from one form to another, but all the different forms of energy in a system always add up to the same total amount of energy, no matter how many energy conversions occur.

Law of Universal Gravitation

The law of universal gravitation states that all objects in the universe attract each other by a force called gravity. The size of the force depends on the masses of the objects and the distance between them.

The first part of the law explains why a bowling ball is much harder to lift than a table-tennis ball. Because the bowling ball has a much larger mass than the table-tennis ball, the amount of gravity between the Earth and the bowling ball is greater than the amount of gravity between the Earth and the table-tennis ball.

The second part of the law explains why a satellite can remain in orbit around the Earth. The satellite is carefully placed at a distance great enough to prevent the Earth's gravity from immediately pulling it down but small enough to prevent it from completely escaping the Earth's gravity and wandering off into space.

Newton's Laws of Motion

Newton's first law of motion states that an object at rest remains at rest and an object in motion remains in motion at constant speed and in a straight line unless acted on by an unbalanced force.

The first part of the law explains why a football will remain on a tee until it is kicked off or until a gust of wind blows it off.

The second part of the law explains why a bike's rider will continue moving forward after the bike tire runs into a crack in the sidewalk and the bike comes to an abrupt stop until gravity and the sidewalk stop the rider.

Newton's second law of motion states that the acceleration of an object depends on the mass of the object and the amount of force applied.

The first part of the law explains why the acceleration of a 4 kg bowling ball will be greater than the acceleration of a 6 kg bowling ball if the same force is applied to both.

The second part of the law explains why the acceleration of a bowling ball will be larger if a larger force is applied to it.

The relationship of acceleration (a) to mass (m) and force (F) can be expressed mathematically by the following equation:

$$\text{acceleration} = \frac{force}{mass} \quad \text{or} \quad a = \frac{F}{m}$$

This equation is often rearranged to the form:

$$\text{force} = \text{mass} \times \text{acceleration}$$
$$\text{or}$$
$$F = m \times a$$

Newton's third law of motion states that whenever one object exerts a force on a second object, the second object exerts an equal and opposite force on the first.

This law explains that a runner is able to move forward because of the equal and opposite force the ground exerts on the runner's foot after each step.

Useful Equations

Average speed

$$\text{Average speed} = \frac{\text{total distance}}{\text{total time}}$$

Example: A bicycle messenger traveled a distance of 136 km in 8 hours. What was the messenger's average speed?

$$\frac{136 \text{ km}}{8 \text{ h}} = 17 \text{ km/h}$$

The messenger's average speed was **17 km/h.**

Average acceleration

$$\frac{\text{Average}}{\text{acceleration}} = \frac{\text{final velocity} - \text{starting velocity}}{\text{time it takes to change velocity}}$$

Example: Calculate the average acceleration of an Olympic 100 m dash sprinter who reaches a velocity of 15 m/s south at the finish line. The race was in a straight line and lasted 10 s.

$$\frac{15 \text{ m/s} - 0 \text{ m/s}}{10 \text{ s}} = 1.5 \text{ m/s/s}$$

The sprinter's average acceleration is **1.5 m/s/s south.**

Net force

Forces in the Same Direction
When forces are in the same direction, add the forces together to determine the net force.

Example: Calculate the net force on a stalled car that is being pushed by two people. One person is pushing with a force of 13 N northwest and the other person is pushing with a force of 8 N in the same direction.

$$13 \text{ N} + 8 \text{ N} = 21 \text{ N}$$

The net force is **21 N northwest.**

Forces in Opposite Directions
When forces are in opposite directions, subtract the smaller force from the larger force to determine the net force.

Net force (cont'd)

Example: Calculate the net force on a rope that is being pulled on each end. One person is pulling on one end of the rope with a force of 12 N south. Another person is pulling on the opposite end of the rope with a force of 7 N north.

$$12 \text{ N} - 7 \text{ N} = 5 \text{ N}$$

The net force is **5 N south.**

Density

$$\text{Density} = \frac{\text{mass}}{\text{volume}}$$

Example: Calculate the density of a sponge with a mass of 10 g and a volume of 40 mL.

$$\frac{10 \text{ g}}{40 \text{ mL}} = 0.25 \text{ g/mL}$$

The density of the sponge is **0.25 g/mL.**

Pressure

Pressure is the force exerted over a given area. The SI unit for pressure is the pascal, which is abbreviated Pa.

$$\text{Pressure} = \frac{\text{force}}{\text{area}}$$

Example: Calculate the pressure of the air in a soccer ball if the air exerts a force of 10 N over an area of 0.5 m².

$$\text{Pressure} = \frac{10 \text{ N}}{0.5 \text{ m}^2} = 20 \text{ N/m}^2 = 20 \text{ Pa}$$

The pressure of the air inside of the soccer ball is **20 Pa.**

Concentration

$$\text{Concentration} = \frac{\text{mass of solute}}{\text{volume of solvent}}$$

Example: Calculate the concentration of a solution in which 10 g of sugar is dissolved in 125 mL of water.

$$\frac{10 \text{ g of sugar}}{125 \text{ mL of water}} = 0.08 \text{ g/mL}$$

The concentration of this solution is **0.08 g/mL.**

Glossary

A

acid precipitation precipitation that contains acids due to air pollution (23)

air mass a large body of air that has similar temperature and moisture throughout (44)

air pressure the measure of the force with which air molecules push on a surface (5)

altitude the height of an object above the Earth's surface (5)

anemometer (AN uh MAHM uht uhr) a device used to measure wind speed (55)

atmosphere a mixture of gases that surrounds a planet, such as Earth (4)

B

barometer an instrument used to measure air pressure (55)

biome a large region characterized by a specific type of climate and the plants and animals that live there (74)

C

cirrus (SIR uhs) **clouds** thin, feathery white clouds found at high altitudes (41)

climate the average weather conditions in an area over a long period of time (68)

cloud a collection of millions of tiny water droplets or ice crystals (40)

condensation the change of state from a gas to a liquid (39)

conduction the transfer of thermal energy from one material to another by direct contact; conduction can also occur within a substance (11)

convection the transfer of thermal energy by the circulation or movement of a liquid or a gas (11)

Coriolis (KOHR ee OH lis) **effect** the curving of moving objects from a straight path due to the Earth's rotation (15)

cumulus (KYOO myoo luhs) **clouds** puffy, white clouds that tend to have flat bottoms (40)

D

deciduous (dee SIJ oo uhs) describes trees that lose their leaves when the weather becomes cold (78)

density the amount of matter in a given space; mass per unit volume (127)

dew point the temperature to which air must cool to be completely saturated (39)

E

elevation the height of an object above sea level; the height of surface landforms above sea level (72)

El Niño periodic change in the location of warm and cool surface waters in the Pacific Ocean (94)

evaporation the change of state from a liquid to a vapor (36)

evergreens trees that keep their leaves year-round (78)

F

front the boundary that forms between two different air masses (46)

G

glacier an enormous mass of moving ice (83)

global warming a rise in average global temperatures (12, 86)

greenhouse effect the natural heating process of a planet, such as the Earth, by which gases in the atmosphere trap thermal energy (12, 86)

H

humidity the amount of water vapor or moisture in the air (37)

hurricane a large, rotating tropical weather system with wind speeds of at least 119 km/h (51)

hypothesis a possible explanation or answer to a question (116)

I

ice age a period during which ice collects in high latitudes and moves toward lower latitudes (83)

isobars lines that connect points of equal air pressure (57)

J

jet streams narrow belts of high-speed winds that blow in the upper troposphere and the lower stratosphere (18)

L

latitude the distance north or south from the equator; measured in degrees (69)

lightning the large electrical discharge that occurs between two oppositely charged surfaces (49)

M

mass the amount of matter that something is made of; its value does not change with the object's location (113)

mesosphere the coldest layer of the atmosphere (8)

meteorology the study of the entire atmosphere (64)

microclimate a small region with unique climatic characteristics (82)

O

observation any use of the senses to gather information (116)

ozone a gas molecule that is made up of three oxygen atoms and that absorbs ultraviolet radiation from the sun (7)

P

polar easterlies wind belts that extend from the poles to 60° latitude in both hemispheres (17)

polar zone the northernmost and southernmost climate zones (80)

precipitation solid or liquid water that falls from the air to the Earth (36, 42)

prevailing winds winds that blow mainly from one direction (71)

primary pollutants pollutants that are put directly into the air by human or natural activity (21)

psychrometer (sie KRAHM uht uhr) an instrument used to measure relative humidity (38)

R

radiation the transfer of energy as electromagnetic waves, such as visible light or infrared waves (10)

relative humidity the amount of moisture the air contains compared with the maximum amount it can hold at a particular temperature (37)

S

scientific method a series of steps that scientists use to answer questions and solve problems (116)

secondary pollutants pollutants that form from chemical reactions that occur when primary pollutants come in contact with other primary pollutants or with naturally occurring substances, such as water vapor (21)

smog a photochemical fog produced by the reaction of sunlight and air pollutants (21)

solar energy energy from the sun (10)

station model a small circle showing the location of a weather station along with a set of symbols and numbers surrounding it that represent weather data (56)

storm surge a local rise in sea level near the shore that is caused by strong winds from a storm, such as a hurricane (53)

stratosphere the atmospheric layer above the troposphere (7)

stratus (STRAT uhs) **clouds** clouds that form in layers (40)

surface current a streamlike movement of water that occurs at or near the surface of the ocean (73)

T

temperate zone the climate zone between the Tropics and the polar zone (78)

temperature a measure of how hot (or cold) something is (8, 54)

thermometer a tool used to measure air temperature (54)

thermosphere the uppermost layer of the atmosphere (8)

thunder the sound that results from the rapid expansion of air along a lightning strike (49)

thunderstorms small, intense weather systems that produce strong winds, heavy rain, lightning, and thunder (48)

tornado a small, rotating column of air that has high wind speeds and low central pressure and that touches the ground (50)

trade winds the winds that blow from 30° latitude to the equator (16)

tropical zone the warm zone located around the equator (75)

troposphere (TROH poh SFIR) the lowest layer of the atmosphere (7)

V

volume the amount of space that something occupies or the amount of space that something contains (125)

W

water cycle the continuous movement of water from water sources into the air, onto land, into and over the ground, and back to the water sources; a cycle that links all of the Earth's solid, liquid, and gaseous water together (36)

weather the condition of the atmosphere at a particular time and place (36, 68)

weather forecast a prediction of future weather conditions over the next 3 to 5 days (54)

westerlies wind belts found in both the Northern and Southern Hemispheres between 30° and 60° latitude (17)

wind moving air (14)

wind energy energy in wind (17)

windsock a device used to measure wind direction (55)

wind vane a device used to measure wind direction (55)

Index

A **boldface** number refers to an illustration on that page.

A

acceleration, average, 127
acid precipitation, 23, **23**
air, 20, **71**. *See also* air pollution; atmosphere
air masses, **44,** 44–47, **47**
air pollution
    car exhaust and, **21,** 22, 87
    cure for, 33
    health effects of, 24, **24**
    indoor, 22
    particulates in, 32
    remediation of, 24–25, 33
    sources of, 22
    types of, 21, **21**
air pressure, 5, **5, 6, 14,** 14–19, **57**
air temperature, 6, **6**
altitude, 5
anemometers, 55, **55**
area
    calculation of, 125
    defined, 125
atmosphere, 25. *See also* air; weather
    composition, 4–5
    heating of, 10–13
    layers of, **6,** 6–9
    pollution of, 20–25, 33. *See also* air pollution
    pressure and temperature, **5,** 5–6, **6, 14,** 14–15, 54
    water vapor in, 7, 36–40, 42
    weather and, 55–56
auroras, 9, **9**
averages, defined, 122

B

barometers, 55
biomes, 74–82, **74–82**
Bradbury, Ray, 65
breezes, **18,** 18–19, **19**
bubonic plague, 94

C

carbon dioxide
    greenhouse effect and, 12, **12,** 86–87
careers in science
    meteorologist, 64, 95
cars
    electric, 33, **33**
    pollution from, 21, **21,** 25, **25**

Celsius scale, 114
chaparrals, **74, 78,** 79, **79**
Clean Air Act of 1970, 24
climate, 68–87. *See also* weather
    changes in, 83–87, 95
    elevation and, 72
    global warming, 12–13, 86–87
    ice ages and, **83,** 83–85
    latitude and, 69, **69**
    microclimates, 82
    models, 95
    mountains and, 72, **72**
    prevailing winds, 71, **71, 72**
    surface currents and, 73, **73**
    volcanoes and, 85, **85**
    zones, 74, **74,** 75, 78, 80
clouds, **40,** 40–42, **41**
    formation of, 40–41
    funnel clouds, 50, **50**
    height of, **41**
    types of, 40–41, **40–41**
cold air masses, 45, **46**
cold fronts, **46, 57**
concentration, calculation of, 127
concept mapping, 112
condensation, **36, 39,** 39–40
conduction, 11, **11**
conifers, 82
conservation of energy, law of, 126
continental drift, 85
continental polar (cP) air masses, **44**
continental tropical (cT) air masses, **44**
controlled experiments, 117
convection
    in the atmosphere, 11, **11**
convection cells, 15, **15**
convection currents, 11, **11**
conversion tables, SI, 113
Coriolis effect, 15, **15**
currents
    climate and, 73, **73**
    convection, 11, **11**
cyclones, 51. *See also* hurricanes

D

deciduous trees, 78
decimals, 123
deforestation, 87, **87**
density
    of air, **71**
    calculation of, 127
deserts
    temperate, **74, 78,** 80, **80**
    tropical, **74, 75,** 76, **76**
dew point, 39
dinosaurs, 32
doldrums, 17

Doppler radar, 56, **56**
dry-bulb thermometer, 38

E

Earth. *See also* atmosphere; maps; plate tectonics
    latitude, **69,** 69–70, **70**
    orbital changes and climate, 84, **84**
electric cars, 33, **33**
electricity
    for cars, 25, 33, **33**
    from wind energy, 17, **17**
elevation, 72
El Niño, 94
energy
    in the atmosphere, 10–11
    transfer of, 10–11, **10–11**
    wind, 17, **17**
Environmental Protection Agency (EPA), 24–25
evaporation, **36**
evergreens, 78
experiments, controlled, 117

F

Fahrenheit–Celsius conversion, 114
Fahrenheit scale, 114, **114**
fingernail growth, 13
first law of motion (Newton's), 126
fog, 40
force(s)
    calculation of, 127
forests, temperate, **74, 78,** 78
fossil fuels,
    problems with, 23–24
fractions, 123–124
fronts, 46–47, **46–47, 57**
funnel clouds, 50, **50**

G

gases
    greenhouse, 12
geography, 72, **72**
glacial periods, 83, **83**
glaciers
    continental drift and, 85
    interglacial periods, 83–84
glaze ice, 42, **42**

global warming, 12–13, 86–87. *See also* climate
global winds, **16,** 16–18
graduated cylinder, 115, **115**

Self-Check Answers

Chapter 1—The Atmosphere

Page 6: As you climb a mountain, the air becomes less dense because there are fewer air molecules. So even though cold air is generally more dense than warm air, it is less dense at higher elevations.

Chapter 2—Understanding Weather

Page 38: Evaporation occurs when liquid water changes into water vapor and returns to the air. Humidity is the amount of water vapor in the air.

Chapter 3—Climate

Page 70: Australia has summer during our winter months, December–February.

Page 76: Because of its dryness, desert soil is poor in organic matter, which fertilizes the soil. Without this natural fertilizer, crops would not be able to grow.

Page 84: The Earth's elliptical orbit causes seasonal differences. When the Earth's orbit is more elliptical, summers are hotter because the Earth is closer to the sun and receives more solar radiation. Winters are cooler because the Earth is farther from the sun and receives less solar radiation.